Structural Engineerin(

10

Matthias Haldimann
Andreas Luible
Mauro Overend

Structural Use of Glass

International Association for Bridge and Structural Engineering IABSE
Association Internationale des Ponts et Charpentes AIPC
Internationale Vereinigung für Brückenbau und Hochbau IVBH

ISBN 978-3-85748-119-2

Publisher:
IABSE-AIPC-IVBH
ETH Zürich
CH-8093 Zürich, Switzerland

Phone: Int. + 41-44-633 2647
Fax: Int. + 41-44-633 1241
E-mail: secretariat@iabse.org
Web: www.iabse.org

Preface

All common applications of architectural glass may be deemed as 'structural'. A small glass pane in a traditional four-edge supported window frame must withstand self-weight, wind-induced pressures, thermal strains and occasional cleaning loads. Unsurprisingly there are several suitable guidelines and adequate rules of thumb that enable appropriate design of these traditional applications.

Recent architectural trends and technological developments have brought about unprecedented opportunities and major changes in the use of glass in buildings. These include the use of large area glass panels; the use of glass in areas traditionally reserved for other materials such as roofs, floors, staircases and partitions; a vast selection of improved glass products; a wide range of novel support and connection details including bolted glass. The consequence of these exciting applications is that glass is often subjected to onerous actions and complex states of stress. Furthermore the glass may now contribute to the integrity of the overall structure and the consequence of failure is considerably greater, such that the glass has a more 'structural' role than the small glass pane in the traditional four-edge supported window. In such applications the traditional rules of thumb are of little assistance as the simplifying assumptions embedded within them no longer hold true and cannot be extrapolated from the specific glass product and simple boundary conditions for which they were devised.

Structural engineers currently have a bewildering array of glass products and configurations to choose from and a wide range of normal and exceptional loading conditions to consider, but very few unified reference texts for undertaking these tasks. This book attempts to redress this issue by providing an overview of the recent developments in this field thereby providing a basis for the understanding of the structural performance and design of glass in buildings.

The book is primarily for structural engineers and researchers who have an interest in structural glass. It draws on topics from many specialist areas such as manufacturing, materials science, fracture mechanics, computational analysis, reliability and forensic engineering and is therefore also relevant to professionals in this field. The level is appropriate for senior undergraduates, post-graduate students, researchers and practising engineers.

The level of interest and the depth of knowledge will vary for instance between the general practising engineer who may be interested in gaining an overview of the general design considerations and specification of glass structures, to the researcher who may be interested in a specific fracture phenomenon in glass. This wide range of interest is accommodated by providing a mix of general and specialist chapters. Furthermore, the text is supplemented by tables of the relevant codes of practice and by an extensive list of references.

The first chapter provides a review of glass production, processing methods and glass products ranging from clear float glass to the more recent developments in switchable glazing. The chapter also provides both a useful listing of glass products as well as the underlying principles that affect the mechanical properties of glass. Chapter 2 provides general guidelines on the analysis and design of glass structures, including advice on actions on glass structures, structural analysis and computational modelling as well as post-breakage behaviour and requirements. Chapter 3 provides an answer to the elusive question 'What is the strength of glass?' by presenting an account of glass fracture mechanisms and formulating useful stochastic failure prediction models for a single flaw and for random surface flaw populations. The chapter also provides an overview of fundamental dynamic fracture mechanics which is useful in glass forensics. Chapter 4 presents a series of quick checks and rules of thumb that are useful for preliminary sizing of structural glass elements. This chapter also reviews the main standards and codes of practice used for the design of structural glass and provides a commentary on the strengths and shortcomings of these standards. Chapter 5 covers the exciting possibilities offered by loading glass in compression and the factors that affect buckling instability in glass. The chapter provides guidelines on the performance and structural design of glass elements that undergo column buckling, lateral torsional buckling and plate buckling. Chapter 6 provides recommendations on how the fracture mechanics formulations presented in the previous chapters may be deployed in practice. This chapter provides guidelines on how to overcome the limitations of laboratory testing and how to obtain statistically relevant and consistent data. Chapter 7 presents a review of connections and support fixings commonly used in glass structures and provides recommendations on good detailing. The chapter also provides useful sizing charts for bolted fixings and reviews the recent developments in enhanced mechanical fixings and adhesive connections. Chapter 8 reviews the current practice and standards used for the design-assisted-by-testing approach, which is particularly relevant for assessing post-breakage and impact performance. This chapter also includes practical advice for undertaking forensic engineering in glass structures.

This book therefore provides a snapshot of the recent developments in the structural use of glass in buildings and draws from the latest developments in practice and research. This was only possible thanks to the contributions from students and colleagues who have kindly donated their work or their time. Their contributions are gratefully acknowledged in the text and the references. The contents of this book have also been greatly enriched by the contributions of several glass experts, who have provided substantial input and advice on specific sections of this book. Their names are listed below and are also shown alongside the headings of the sections they contributed in.

Benjamin BEER	Werner Sobek Stuttgart, Germany
Lucio BLANDINI, Dr.	Werner Sobek Stuttgart, Germany
Mick EEKHOUT, Prof. Dr.	Octatube, Delft, The Netherlands
Christoph HAAS	Ernst Basler + Partner AG, Zürich, Switzerland
Iris MANIATIS, Dr.	University of the German Federal Armed Forces Munich, Germany
Jürgen NEUGEBAUER, Dr.	University of Applied Sciences, FH Joanneum, Graz, Austria
Jens SCHNEIDER, Prof. Dr.	Frankfurt University of Applied Sciences, Germany
Werner SOBEK, Prof. Dr.	Werner Sobek Stuttgart, Germany
Geralt SIEBERT, Prof. Dr.	University of the German Federal Armed Forces Munich, Germany
Ronald VISSER	Octatube, Delft, The Netherlands
Frank WELLERSHOFF, Dr.	Permasteelisa Central Europe GmbH, Würzburg, Germany

Finally a special word of thanks goes to the IABSE Structural Engineering Documents Editorial Board, especially to Geoff Taplin, and to their reviewers Chris Jofeh and Geralt Siebert, who kindly reviewed the drafts of this book and helped to improve it with their valuable feedback.

Berne, Basel and Nottingham/January 2008

Dr. Matthias Haldimann
Dr. Andreas Luible
Dr. Mauro Overend

Contents

Chapter 1

Material

This text has been compiled in collaboration with the following experts:
Prof. Dr. Jens Schneider

1.1 Production

1.1.1 Production of flat glass

Figure 1.1 gives an overview of the most common glass production processes, processing methods and glass products. The main production steps are always similar: melting at 1600–1800 °C, forming at 800–1600 °C and cooling at 100–800 °C.

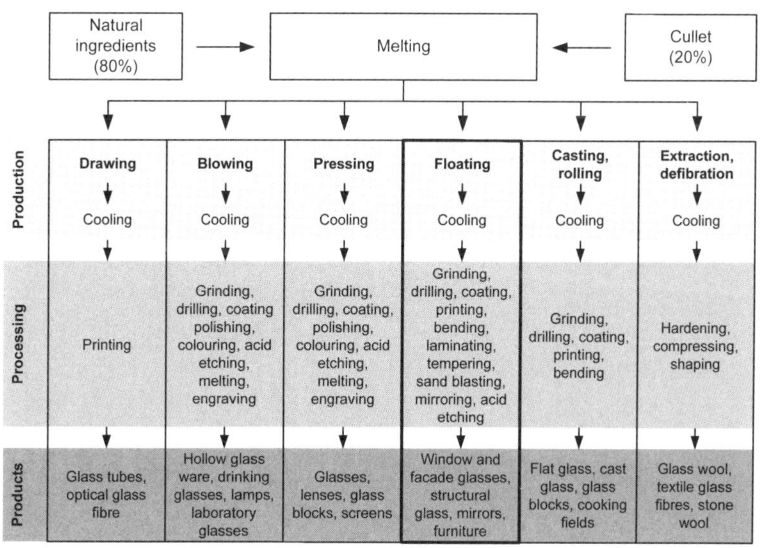

Figure 1.1: Glass production processes and products overview.

Currently the float process is the most popular primary manufacturing process and accounts for about 90% of today's flat glass production worldwide. The major advantages of this production process, introduced commercially by the Pilkington Brothers in 1959, are its low cost, its wide availability, the superior optical quality of the glass and the large size of panes that can be reliably produced. The mass production process together with many post-processing and refinement technologies invented or improved over the last 50 years (see Section 1.3) have made glass cheap enough to allow it to be used extensively in the construction industry and arguably to become 'the most important material in architecture' (Le Corbusier). Within the last two decades, further progress in the field of refinement technologies (tempering, laminating) aided by structural analysis methods (e. g. finite element method) have enabled glass to be used for structural building elements.

Float glass is made in large manufacturing plants that operate continuously 24 hours a day, 365 days a year. The production process is shown schematically in *Figure 1.2*. The raw materials are melted in a furnace at temperatures of up to 1550 °C. The molten glass is then poured continuously at approximately 1000 °C on to a shallow pool of molten tin whose oxidation is prevented by an inert atmosphere consisting of hydrogen and nitrogen. Tin is used because of the large temperature range of its liquid physical state (232–2270 °C) and its high specific weight in comparison with glass. The glass floats on the tin and spreads outwards forming a smooth flat surface at an equilibrium thickness of 6 to 7 mm, which is gradually cooled and drawn on to rollers, before entering a long oven, called a lehr, at around 600 °C. The glass thickness can be controlled within a range of 2 to 25 mm by adjusting the speed of the rollers. Reducing the speed increases glass thickness and vice versa. The annealing lehr slowly cools the glass to prevent residual stresses being induced within the glass. After annealing, the float glass is inspected by automated machines to ensure that obvious visual defects and imperfections are removed during cutting. The glass is cut to a typical size of 3.21 m × 6.00 m before being stored. Any unwanted or broken glass is collected and fed back into the furnace to reduce waste. At some float plants, so called on-line coatings (hard coatings) can be applied to the hot glass surface during manufacture.

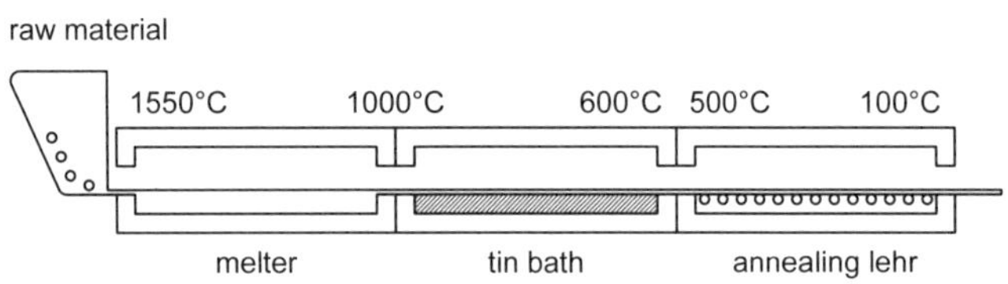

Figure 1.2: Production process for float glass.

As a consequence of this production process, the two faces of glass sheets are not completely identical. On the tin side, some diffusion of tin atoms into the glass surface occurs. This may have an influence on the behaviour of the surface when it is glued [241]. The mechanical strength of the tin side has been found to be marginally lower than that of the air side. This is not attributed to the diffused tin atoms but to the contact of the tin side with the transport rollers in the cooling area. These rollers cause some surface flaws that reduce the strength [302]. This interpretation is supported by the fact that the strength of intentionally damaged glass specimens has been found to be independent of the glass side [184]. The tin side can be detected thanks to its bluish fluorescence when exposed to ultraviolet radiation.

1.1.2 Production of cast glass and glass profiles

The cast process is an older production process for flat glass. The molten glass is poured continuously between metal rollers to produce glass with a controlled thickness (*Figure 1.3*). The rollers may be engraved to give the glass a surface design or texture and produce patterned glass. In a simple modification of the process, a steel wire mesh can be sandwiched between two separate ribbons of glass to produce wired glass. Cast glass (also called rolled glass) was first produced in 1870, wired glass in 1898 [225]. Annealing is performed in a way similar to the float process.

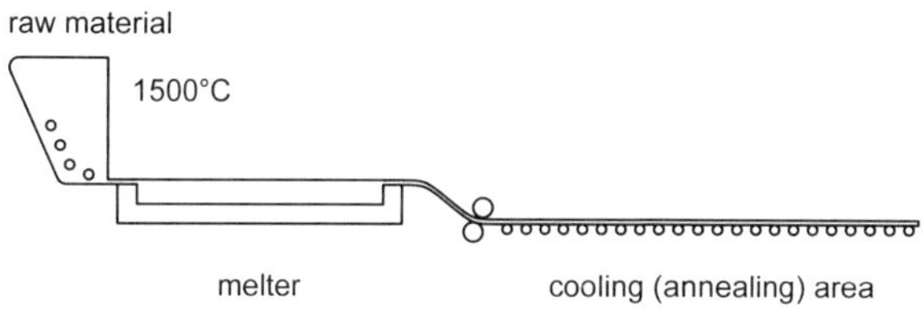

Figure 1.3:
Production process for cast glass and glass profiles.

Cast glass is usually not transparent, but translucent. Flat surfaces must be polished to obtain a truly clear glass. Wired glass was formerly known as 'safety glass' or 'fire protection glass' as the wire mesh keeps most of the glass pieces together after breakage. But the risk of injuries by sharp splinters remains high. Today, laminated glasses and special fire protection glasses with a much better safety performance are preferred to wired glass.

The production of glass profiles is currently limited to U-shaped profiles (or channel shaped glass) and circular hollow sections (tubes). U-profiles are produced on the basis of the cast process, using additional rollers to bend the edges of the glass. U-profiles can also be formed using wired glass. While glass profiles have traditionally been mainly used as a substitute of windows in industrial structures, they have been rediscovered for modern façades in recent years.

Traditionally, glass tubes have mainly been produced for the chemical industry. The most common production process is the Danner process, named after the American engineer Edward Danner, who developed this process in 1912. In the Danner process, the glass flow falls onto a rotating, slightly downward pointing mandrel. Air is blown down a shaft through the middle of the mandrel, thus creating a hollow space in the glass as it is drawn off the end of the mandrel by a tractor mechanism. The diameter and thickness of the glass tubing can be controlled by regulating the strength of the air flow through the mandrel and the speed of the drawing machine. The process allows for wall thicknesses of up to 10 mm only. The more recent centrifuging process allows the production of large sections and non-rotationally symmetrical items by spinning, but is expensive [348]. In this process, molten glass is fed into a steel mould that rotates at the required speed. At high speeds, the glass can assume almost cylindrical shapes. When the glass has cooled sufficiently, rotation stops and the glass is removed.

1.1.3 Relevant standards

Table 1.1 gives an overview of important European and US standards for basic glass products. For standards on processed glass products, see *Table 1.9*.

Table 1.1: Important standards for basic glass products (shortened titles).

Standard	Title
EN 572-1:2004 [148]	Basic soda lime silicate glass products – Part 1: Definitions and general physical and mechanical properties
EN 572-2:2004 [149]	Basic soda lime silicate glass products – Part 2: Float glass
EN 572-3:2004 [150]	Basic soda lime silicate glass products – Part 3: Polished wire glass
EN 572-4:2004 [151]	Basic soda lime silicate glass products – Part 4: Drawn sheet glass
EN 572-5:2004 [152]	Basic soda lime silicate glass products – Part 5: Patterned glass
EN 572-6:2004 [153]	Basic soda lime silicate glass products – Part 6: Wired patterned glass
EN 572-7:2004 [154]	Basic soda lime silicate glass products – Part 7: Wired or unwired channel shaped glass
EN 572-8:2004 [155]	Basic soda lime silicate glass products – Part 8: Supplied and final cut sizes
EN 572-9:2004 [156]	Basic soda lime silicate glass products – Part 9: Evaluation of conformity/Product standard
ASTM C 1036-2001 [11]	Standard Specification for Flat Glass
EN 1748-1-1:2004 [129]	Special basic products – Borosilicate glasses – Part 1-1: Definitions and general physical and mechanical properties
EN 1748-1-2:2004 [130]	Special basic products – Borosilicate glasses – Part 1-2: Evaluation of conformity/ Product standard
EN 1748-2-1:2004 [131]	Special basic products – Glass ceramics – Part 2-1 Definitions and general physical and mechanical properties
EN 1748-2-2:2004 [132]	Special basic products – Glass ceramics – Part 2-2: Evaluation of conformity/ Product standard
EN 1051-1:2003 [95]	Glass blocks and glass paver units – Part 1: Definitions and description
EN 1051-2:2003 [96]	Glass blocks and glass paver units – Part 2: Evaluation of conformity

1.2 Material properties

1.2.1 Composition and chemical properties

A *glass* is an inorganic product of fusion, which has been cooled to a rigid condition without crystallization. The term therefore applies to all noncrystalline solids showing a glass transition[1]. Most of the glass used in construction is soda lime silica glass (SLSG). For some special applications (e. g. fire protection glazing, heat resistant glazing), borosilicate glass (BSG) is used. The latter offers very high resistance to temperature changes as well as a very high hydrolytic and acid resistance. *Table 1.2* gives the chemical composition of these two glass types according to current European standards. In contrast to most other materials, glasses do not consist of a geometrically regular network of crystals, but of an irregular network of silicon and oxygen atoms with alkaline parts in between (*Figure 1.4*). The chemical composition has an important influence on the viscosity, the melting temperature T_S and the thermal expansion coefficient α_T of glass. While the melting temperature is about 1 710 °C for pure silica oxide, it drops to 1 300–1 600 °C through the addition of alkali. The thermal expansion coefficient is about $0.5 \cdot 10^{-6}\,K^{-1}$ for pure silica glass and $9 \cdot 10^{-6}\,K^{-1}$ for soda lime silica glass.

During the cooling of the liquid glass, its viscosity increases constantly until solidification at about 10^{14} Pa s. The temperature at solidification is called glass transition temperature T_g and is about 530 °C for SLSG. In contrast to crystalline materials, the transition between liquid and solid states does not take place at one precise temperature but over a certain temperature range (*Figure 1.5*, *Table 1.3*).

[1] A pseudo second order phase transition in which a supercooled melt yields, on cooling, a glassy structure and properties similar to those of crystalline materials.

		Soda lime silica glass	Borosilicate glass
Silica sand	SiO_2	69–74%	70–87%
Lime (calcium oxide)	CaO	5–14%	–
Soda	Na_2O	10–16%	0–8%
Boron-oxide	B_2O_3	–	7–15%
Potassium oxide	K_2O	–	0–8%
Magnesia	MgO	0–6%	–
Alumina	Al_2O_3	0–3%	0–8%
Others		0–5%	0–8%

Table 1.2:
Chemical composition of soda lime silica glass and borosilicate glass; indicatory values (mass %) according to [148] and [129].

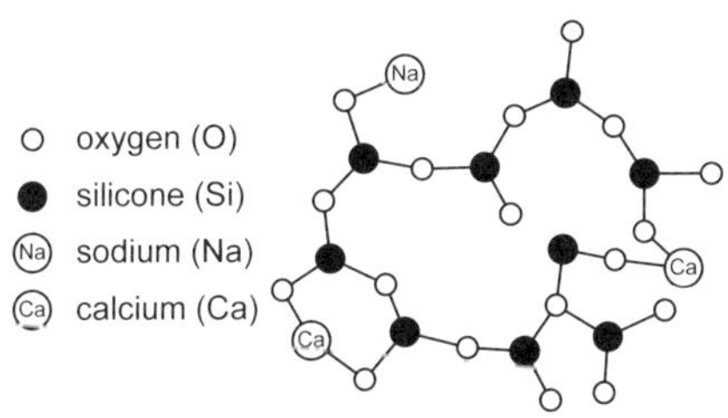

Figure 1.4:
Schematic view of the irregular network of a soda lime silica glass.

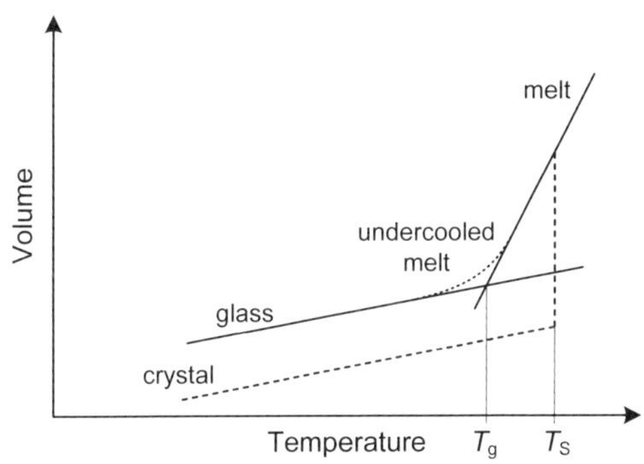

Figure 1.5:
Schematic comparison of the volume's dependence on temperature for a glass and a crystalline material.

Viscosity	State	Temperature SLSG	Temperature BSG
(dPa s)		(°C)	(°C)
10^5	Working point	1040	1280
$10^{8.6}$	Softening point	720	830
10^{14}	Annealing point	540	570
$10^{14.3}$	Transition temperature T_g	530	560
$10^{15.5}$	Strain point	506	530

Table 1.3:
Typical viscosities and corresponding temperatures for soda lime silica glass (SLSG) and borosilicate glass (BSG).

Unlike most solids, the electrons in glass molecules are strictly confined to particular energy levels. Since this means that the molecules cannot alternate between different states of excitement by absorbing radiation in the bandwidths of visible and near infrared, they do not absorb or dissipate those forms of radiant energy. Instead, the energy passes straight through the molecules as if they were not there. However, due to unavoidable impurities in the soda-lime-silica mix, typical window glass does absorb some radiation that might otherwise pass through (cf. Section 1.2.2). Small amounts of iron oxides are responsible for the characteristic greenish colour of soda lime silica glass (e. g. Fe^{2+}: blue-green; Fe^{3+}: yellow-brown). Extra-clear glass, so-called low iron glass, which has a reduced iron oxide content in order to lessen the green tinge, is commercially available.

One of the most important properties of glass is its excellent chemical resistance to many aggressive substances, which explains its popularity in the chemical industry and makes glass one of the most durable materials in construction (*Table 1.4*).

Table 1.4: Qualitative overview of the chemical resistance of soda lime silica glass.

Substance	**Resistance**
Non-oxidant and oxidant acids	+
SiO_2-solving acids	0/−
Salt	+
Water	+
Non-oxidant and oxidant alkalis	0/−
Aliphatic, aromatic and chlorinated hydrocarbons	+
Alcohol	+
Ester	+
Ketones	+
Oil and Fat	+

+: resistant, 0: partly resistant, −: not resistant.

1.2.2 Physical properties

The most important physical properties of SLSG and BSG are summarized in *Table 1.5*. Optical properties depend on the glass thickness, the chemical composition and the applied coatings. The most evident property is the very high transparency within the visible range of wavelengths ($\lambda \approx 380$–$750\,\text{nm}$). Whilst the exact profiles of the non-transmitted (i. e. absorbed and reflected) radiation spectrum varies between different types of glass, they are usually in the wavelengths outside the visible and near infrared band (*Figure 1.6*). Due to interaction with O_2-ions in the glass, a large percentage of UV radiation is absorbed. Long-wave infrared radiation ($\lambda > 5\,000\,\text{nm}$) is blocked because it is absorbed by Si-O-groups. This is at the origin of the greenhouse effect: visual light passes through the glass and heats up the interior, while emitted long-wave thermal radiation is unable to escape. With its refractive index of about 1.5, the reflection of visual light by common SLSG is 4% per surface which gives a total of 8% for a glass pane. This reduces transparency but can be avoided by applying special coatings.

At room temperature, the dynamic viscosity of glass is about 10^{20} dPa s. (For comparison, the viscosity of water is 10^{-1} dPa s and of honey, 10^{5} dPa s.) Given this extremely high viscosity at room temperature, it would take more than the earth's age for 'flow'

Table 1.5: Physical properties of soda lime silica glass (SLSG) and borosilicate glass (BSG) [129, 148].

			Soda lime silica glass	Borosilicate glass
Density	ρ	kg/m^3	2 500	2 200–2 500
Knoop hardness	$HK_{0,1/20}$	GPa	6	4.5–6
Young's modulus	E	MPa	70 000	60 000–70 000
Poisson's ratio	ν	–	0.23*	0.2
Coefficient of thermal expansion†	α_T	$10^{-6}\,K^{-1}$	9	Class 1: 3.1–4.0 Class 2: 4.1–5.0 Class 3: 5.1–6.0
Specific thermal capacity	c_p	$J\,kg^{-1}\,K^{-1}$	720	800
Thermal conductivity	λ	$W\,m^{-1}\,K^{-1}$	1	1
Average refractive index within the visible spectrum‡	n	–	1.52§	1.5
Emissivity (corrected¶)	ε	–	0.837	0.837

* EN 572-1:2004 [148] gives 0.2. In research and application, values between 0.22 and 0.24 are commonly used.
† Mean between 20 °C and 300 °C.
‡ The refractive index is a constant for a given glazing material, but depends on the wavelength. The variation being small within the visible spectrum, a single value provides sufficient accuracy.
§ EN 572-1:2004 [148] gives a rounded value of 1.50.
¶ For detailed information on the determination of this value see EN 673:1997 [157].

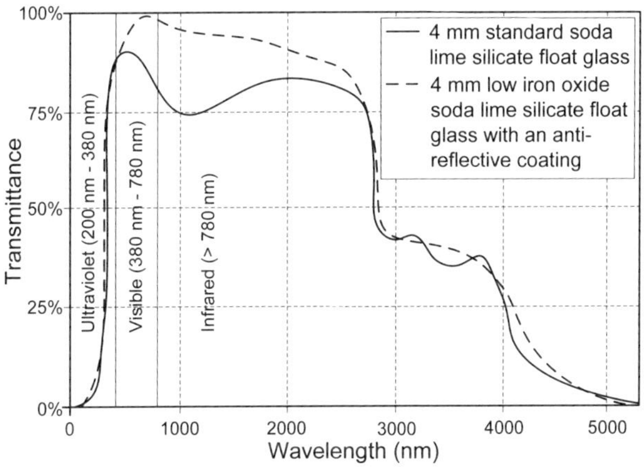

Figure 1.6: Transmittance as a function of wavelength for a typical soda lime silica glass and a low-iron glass with anti-reflective coating.

effects to become visible to the naked eye. Although the notion of flowing glass has been repeatedly propagated, 'flow' of the glass is however not the cause of window glasses in old churches being often thicker at the bottom than at the top.[2]

[2]The likely source of this observation is that when panes of glass were commonly made by glassblowers, the technique used was to spin molten glass so as to create a round, mostly flat and even plate (crown glass process). This plate was then cut to fit a window. The pieces were not, however, absolutely flat; the edges of the disk would be thicker because of centripetal force relaxation. When actually installed in a window frame, the glass would be placed thicker side down for the sake of stability and visual sparkle. Occasionally such glass has been found thinner side down or on either side of the windows edge, as would be caused by carelessness at the time of installation. [347]

Glass shows an almost perfectly elastic, isotropic behaviour and exhibits brittle fracture. It does not yield plastically, which is why local stress concentrations are not reduced through stress redistribution as is the case for other construction materials like steel. The theoretical tensile strength (based on molecular forces) of glass is exceptionally high and may reach 32 GPa. It is, however, of no practical relevance for structural applications. The actual tensile strength, the relevant property for engineering, is much lower. The reason is that as with all brittle materials, the tensile strength of glass depends very much on mechanical flaws on the surface. Such flaws are not necessarily visible to the naked eye. While the surface of glass panes generally contains a large number of relatively severe flaws, the surface of glass fibres contains less and less deep surface flaws. This explains the much higher strength of glass fibres when compared to glass panes. *Figure 1.7* gives a rough overview of typical strength values for various flaw depths.

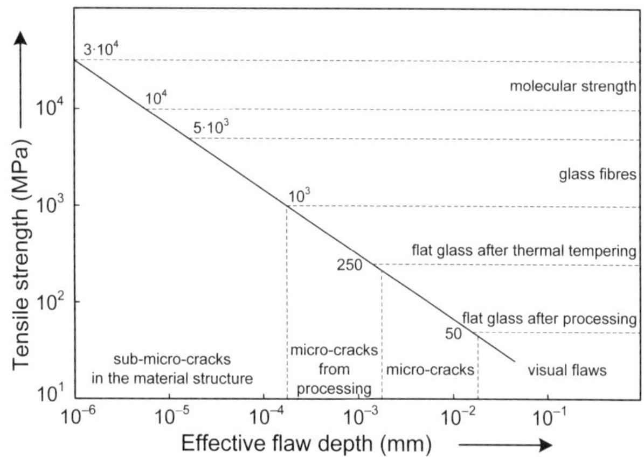

Figure 1.7:
Typical short-term strengths as a function of the flaw depth (adapted from [274]).

A glass element fails as soon as the stress intensity due to tensile stress at the tip of one flaw reaches its critical value. Flaws grow with time when loaded, the crack velocity being a function of several parameters and extremely variable. This is discussed in detail in Chapter 3. For the moment, it shall only be pointed out that the tensile strength of glass is not a material constant, but it depends on many aspects, in particular on the condition of the surface, the size of the glass element, the action history (intensity and duration), the residual stress and the environmental conditions. The higher the load, the longer the load duration and the deeper the initial surface flaw, the lower the effective tensile strength.

As surface flaws do not grow or fail when in compression, the compressive strength of glass is much larger than the tensile strength. Nevertheless, the compressive strength is irrelevant for virtually all structural applications. Tensile stresses develop because of buckling in the case of stability problems and because of the Poisson's ratio effect at load introduction points. In both cases, an element's tensile strength is exceeded long before it is loaded to its compressive strength.

1.3 Processing and glass products

1.3.1 Introduction

Once manufactured, flat glass is often processed further to produce glass products of the shape, performance and appearance that are required to meet particular needs. This secondary processing may include:

- cutting to remove edge damage and to produce the desired pane shape and size
- edge working (arrissing, grinding, polishing)
- hole drilling
- curving
- application of coatings
- thermal treatment to get heat strengthened or fully tempered glass (tempering)
- heat soaking to reduce the potential for nickel sulfide-induced breakage of fully tempered glass in use
- laminating for enhanced post-breakage performance, safety on impact, bullet resistance, fire resistance or acoustic insulation
- surface modification processes for decoration, shading or privacy
- insulating glass unit assembly to reduce heat loss and, if suitably configured, to reduce solar gain and enhance acoustic performance.

The term *glass pane* will hereinafter be used to refer to a single pane of sheet glass. A glass pane may be used as a monolithic glass or it may be part of an insulating glass unit, a laminated glass or some other glass assembly (*Figure 1.8*). *Glass unit* is a generic term for any of these.

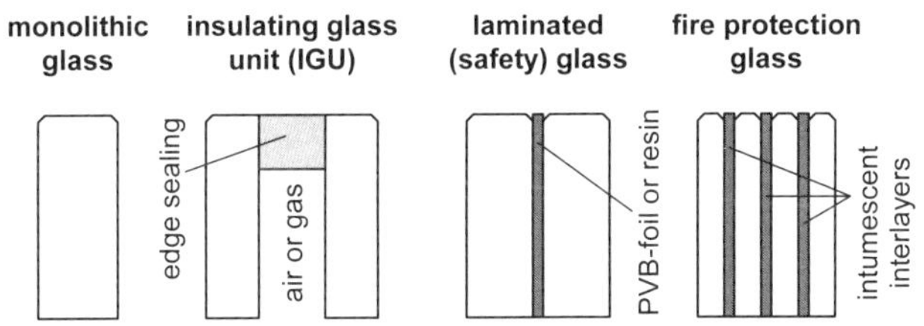

Figure 1.8: Basic types of glass units.

The following sections give detailed information on the most important glass products and processing methods used in construction.

1.3.2 Tempering of glass

Principle and main effects

For structural glass applications, tempering (heat treatment) is the most important processing method. The idea is to create a favourable residual stress field featuring tensile stresses in the core of the glass and compressive stresses on and near the surfaces. The glass core does not contain flaws and therefore offers good resistance to tensile stress. The unavoidable flaws on the glass surface can only grow if they are exposed to an effective tensile stress. As long as the tensile surface stress due to actions is smaller than the

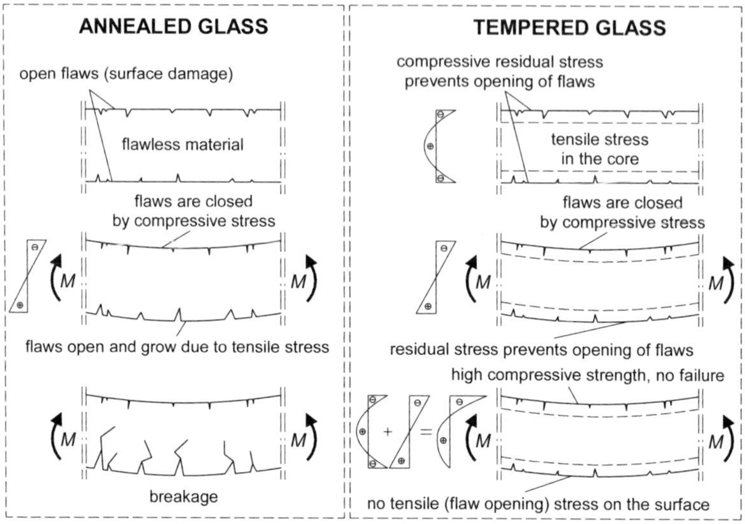

Figure 1.9: The principle of glass tempering (adapted from [302]).

residual compressive stress, there is no such effective tensile stress and consequently no crack growth (*Figure 1.9*).

The fracture pattern is a function of the energy stored in the glass, i. e. of the residual stress and the stress due to loads. As an example, *Figure 1.10* shows the fracture pattern of specimens loaded in a coaxial double ring test setup. *Fully tempered glass* has the highest residual stress level and usually breaks into small, relatively harmless dice of about 100 mm^2. This fracture pattern is why fully tempered glass is also called *'safety glass'*. The term may, however, be misleading. When falling from a height of several meters, even small glass dice can cause serious injury. While fully tempered glass has the highest structural capacity of all glass types, its post-failure performance is poor due to the tiny fragments. *Heat strengthened glass* provides an interesting compromise between fairly good structural performance and a sufficiently large fragmentation pattern for good post-failure performance. *Annealed glass* is standard float glass without any tempering. It normally breaks into large fragments. If, however, it is exposed to high (especially in-plane) loads, the elastic energy stored in the material due to elastic deformation can lead to a fracture pattern similar to heat-treated glass.

On an international level, no specific terminology for the different glass types has to date gained universal acceptance. In the present document, the terms from ASTM E 1300-04 [22] are used (*Table 1.6*). They are widely used and tend, in the opinion of the authors, to be less susceptible to misunderstandings than others.

Fully tempered glass

During the thermal tempering process (*Figure 1.11*), float glass is heated to approximately 620 – 675 °C (approximately 100 °C above the glass transition temperature) in a furnace and then quenched (cooled rapidly) by jets of cold air. This has the effect of cooling and solidifying first the surface and then the interior of the glass (*Figure 1.12*). Within the first seconds, the cooling process results in tensile stresses on the surface and compressive

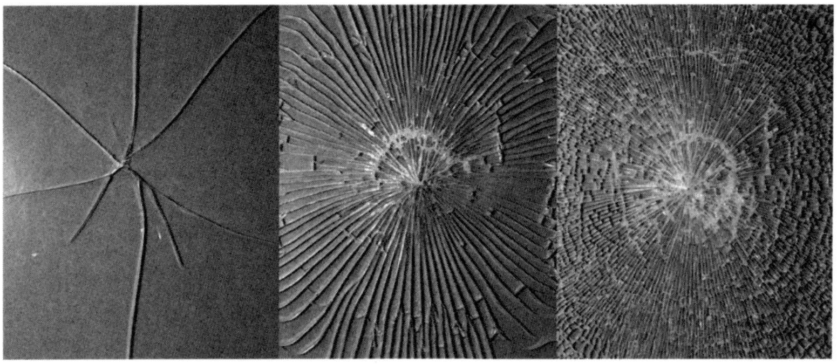

Figure 1.10: Comparison of the fracture pattern: annealed glass (left), heat strengthened glass (middle), fully tempered glass (right).

Table 1.6: Glass type terminology overview.

Level of residual surface compression	Terminology in the present document	Other frequently used terms
(Almost) none	Annealed glass (ANG)	Float glass
Medium	Heat strengthened glass (HSG)	Partly toughened glass;
High	Fully tempered glass (FTG)	Tempered glass; (thermally) toughened glass
Unspecified (HSG or FTG)	Heat-treated glass	

stresses in the interior. As the glass is viscous in this temperature range, the tensile stresses can relax rapidly. If the starting temperature is too low, the relaxation cannot take place and the tensile stresses may cause the glass to shatter in the furnace. As soon as the temperature on the glass surface falls below T_g (approximately 525 °C), the glass solidifies and relaxation stops immediately. The temperature distribution is approximately parabolic, the interior being hotter at this stage. Finally, the interior cools as well. As its thermal shrinkage is resisted by the already solid surface, the cooling leads to the characteristic residual stress field with the surfaces being in compression and the interior in tension. To obtain an optimal result with maximum temper stress, the process has to be managed so that the surface solidifies exactly at the moment when the maximum temperature difference occurs and the initial tensile stress has relaxed. Borosilicate glass is difficult to temper by high air pressure or even by quenching in liquids because of its low thermal expansion coefficient.

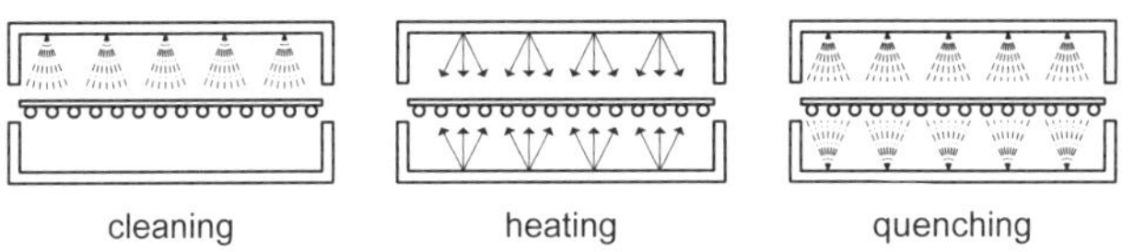

Figure 1.11: Tempering process.

The typical residual compressive surface stress varies between 80 MPa and 170 MPa for fully tempered soda lime silica glass. In ASTM C 1048-04 [12], it is required to have either a minimum surface compression of 69 MPa (10 000 psi) or an edge compression of not less than 67 MPa (9 700 psi). In European standards, the fragmentation count, the

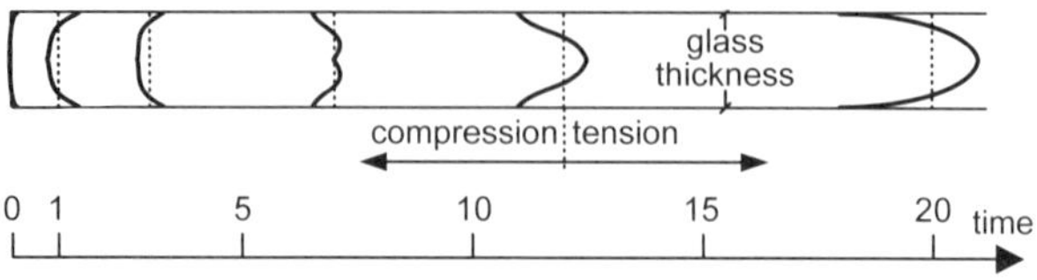

Figure 1.12: Transient stress field during the tempering process.

maximum fragment size and the minimum fracture strength in four point bending tests is specified [101, 102].

Fairly accurate numerical modelling of the tempering process is possible [42, 61, 62, 63, 64, 237, 297]. This is especially helpful to estimate tempering stresses for more complex geometries like boreholes. The most important parameters of the tempering process are the glass thickness, the thermal expansion coefficient of the glass and the heat transfer coefficient between glass and air. Particularly the heat transfer coefficient is often difficult to estimate. It depends on the quenching (jet geometry, roller influence, air pressure, air temperature, etc.) and is therefore quite variable for different glass manufacturers.

Heat strengthened glass

Heat strengthened glass is produced using the same process as for fully tempered glass, but with a lower cooling rate. The residual stress and therefore the tensile strength is lower. The fracture pattern of heat strengthened glass is similar to annealed glass, with much bigger fragments than for fully tempered glass. Used in laminated glass elements, this large fracture pattern results in a significant remaining load-bearing capacity after failure.

As the stress gradient depends on the glass thickness and the glass must be cooled down slowly, thick glasses (> 12 mm) cannot be heat strengthened using the normal tempering process.

The typical residual compressive surface stress varies between 40 and 80 MPa for heat strengthened glass. ASTM C 1048-04 [12] requires that heat strengthened glass has a residual compressive surface stress between 24 MPa (3 500 psi) and 52 MPa (7 500 psi). In European standards, the fragmentation count and the maximum fragment size is specified [133, 134].

Chemical tempering

Chemical tempering is an alternative tempering process that does not involve thermic effects and produces a different residual stress profile. Cutting or drilling remains possible, even after tempering. In structural applications, chemical tempering is extremely rare. It is used for special geometries where usual tempering processes cannot be applied, e. g. glasses with narrow bending angles. The process is based on the exchange of sodium ions in the glass surface by potassium ions, which are about 30% bigger. Only a very thin zone at the glass surface is affected (*Figure 1.13*). The actual depth of the compression zone is time dependent (about 20 µm in 24 hours) [348]. If surface flaws are deeper than the compression zone, their tip is in the zone of tensile stress and subcritical crack growth occurs without external load. This phenomenon, known as *self-fatigue*, can cause spontaneous failure, even of glass elements that have never been exposed to external loads. For a fracture mechanics investigation, see [27]. An improved chemical tempering process is currently being developed, see e. g. [2, 304, 305]. While the scatter of the strength can be reduced, the problem of self fatigue persists and the process is expensive.

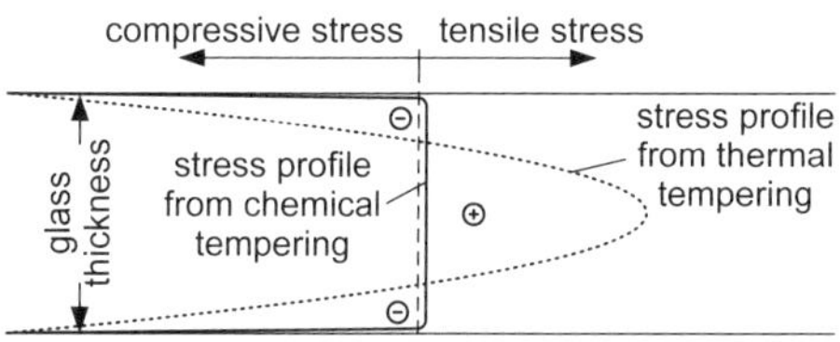

Figure 1.13:
Comparison of the stress profiles obtained by thermal and chemical tempering.

Tolerances and practical aspects

An attempt to work heat-treated glass usually causes it to shatter immediately. Any cutting, drilling or grinding must therefore be carried out *before* the glass is tempered.

The heating of the glass to more than the transition temperature and the fixing in the furnace causes some deformation. It depends on the furnace and the glass thickness, but generally increases with increasing aspect ratio of a glass element. This can limit the feasible slenderness of glass beams. Furthermore, geometric tolerances are considerably higher than those of annealed glass. In particular, edges and holes in laminated glass elements made of heat-treated glass are generally not flush. This cannot be corrected by grinding (because grinding would cause the glass to shatter, see above) and must therefore be accounted for by well thought-out details and connections. Finally, the deformation often reduces the optical quality of heat-treated glass.

Specialized glass processing firms are able to temper bent glasses, but various limitations on radii and dimensions may apply.

Nickel sulfide-induced spontaneous failure

Fully tempered glass elements have a small but not negligible risk of breaking spontaneously within a few years of production. At the origin of such spontaneous failures are nickel sulfide (NiS) inclusions (*Figure 1.14*) that cannot be avoided completely during production. Under the influence of temperature, such NiS particles can increase in volume by about 4% due to a phase change. This expansion in combination with the high tensile stresses in the glass core due to thermal tempering can cause spontaneous failure.

The risk of spontaneous failure due to inclusions can be significantly reduced, but not totally eliminated[3], by the heat-soak test. This test consists of slowly heating up the glass and maintaining a certain temperature for several hours. This accelerates the phase change, and glass elements containing dangerous inclusions fail during the test. Depending on the location, client and glass processor involved, the heat-soak test is performed according to DIN 18516-4:1990 [83], EN 14179-1:2005 [123] or the German building regulation BRL-A 2005 [46]. All three regulations specify a holding temperature of $290 \pm 10\,^{\circ}$C. The duration of the holding period is 8 hours according to DIN 18516-4:1990 [83], 4 hours according to BRL-A 2005 [46] and 2 hours according to EN 14179-1:2005 [123].

[3] According to EN 14179-1:2005 [123], there is at most one failure in 400 t of heat soaked glass.

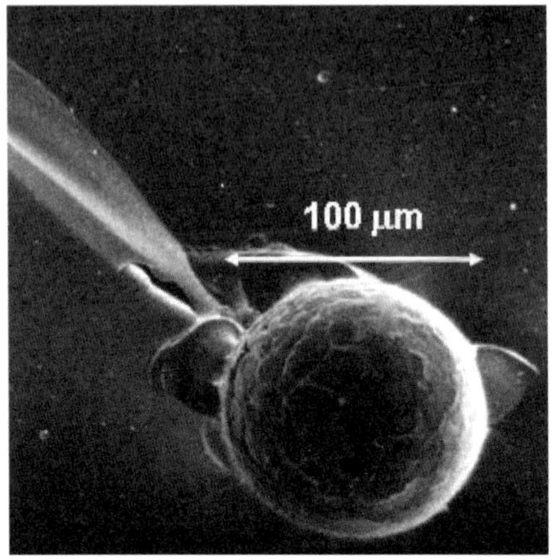

Figure 1.14:
Microscopic image of a nickel-sulfide inclusion in fully tempered glass (courtesy of MPA Darmstadt, Germany).

1.3.3 Laminated glass

Laminated glass consists of two or more panes bonded together by some transparent plastic interlayer. The glass panes may be equal or unequal in thickness and may be the same or different in heat treatment. The most common lamination process is autoclaving at approximately 140 °C. The heat and the pressure of up to 14 bar ensure that there are no air inclusions between the glass and the interlayer.

Laminated glass is of major interest in structural applications. Even though tempering reduces the time dependence of the strength and improves the structural capacity of glass, it is still a brittle material. Lamination of a transparent plastic film between two or more flat glass panes enables significant improvement in the post-breakage behaviour: after breakage, the glass fragments adhere to the film so that a certain remaining structural capacity is obtained as the glass fragments 'arch' or lock in place. This capacity depends on the fragmentation of the glass and increases with increasing fragment size (*Figure 1.15*). Therefore, laminated glass elements achieve a particularly high remaining structural capacity when made from annealed or heat strengthened glass that breaks into large fragments. The post-breakage behaviour furthermore depends on the interlayer material.

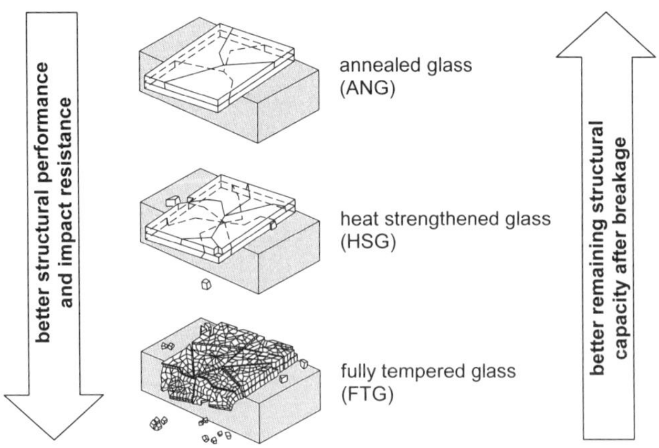

Figure 1.15:
Post breakage behaviour of laminated glass made of different glass types (adapted from [302]).

The most common interlayer material is polyvinyl butyral (PVB). Because PVB blocks UV radiation almost completely, PVB foils are sometimes also called UV-protection foils. The nominal thickness of a single PVB foil is 0.38 mm. Normally, two (0.76 mm) or four (1.52 mm) foils form one PVB interlayer. For heat-treated or curved glasses, up to six may be appropriate to compensate for the unevenness of the glass panels due to tempering (see Section 1.3.2). PVB is a viscoelastic material, i. e. its physical properties depend strongly on the temperature and the load duration. At room temperature, PVB is comparatively soft with an elongation at breakage of more than 200%. At temperatures well below 0 °C and for short loading times, PVB is in general able to transfer the full shear stress from one pane of glass to another. For higher temperatures and long loading times, the shear transfer is greatly reduced. *Table 1.7* gives typical properties of PVB.

Density	ρ	kg/m^3	1 070
Shear modulus	G	GPa	0–4
Poisson's ratio	ν	–	≈0.50
Coefficient of thermal expansion	a_T	K^{-1}	$80 \cdot 10^{-6}$
Tensile strength	f_t	MPa	≥20
Elongation at failure	ε_t	%	≥300

Table 1.7:
Typical material properties of PVB.

Alternative transparent interlayer materials have recently been developed with the aim of achieving higher stiffness, temperature resistance, tensile strength or resistance to tearing. A well known example is DuPont's SentryGlass Plus [40, 93, 276]. However the high stiffness can make the lamination of such interlayers difficult.

In addition to the transparent interlayers, coloured or printed ones are also available. Other materials, i. e. transparent 'cold poured' resins with 1 to 4 mm layer thickness, are sometimes used to achieve special properties like sound insulation or to include functional components like solar cells or light emitting diodes (LEDs).

Fire protection glass is laminated glass with one or more special transparent intumescent interlayer(s). When exposed to fire, the pane facing the flames fractures but remains in place and the interlayers foam up to form an opaque insulating shield that blocks the heat of the blaze.

Bullet-resistant and blast-resistant glasses are laminated glasses using various impact energy absorbing interlayers. In some applications one or more of the sandwiched glass panes may be replaced by a polycarbonate pane.

1.3.4 Insulating glass units (IGU)

An insulating glass unit (IGU) is a multi-glass combination consisting of two or more panes enclosing a hermetically sealed air space (*Figure 1.16*). The most important function of IGUs is to reduce thermal losses. Besides the advantage of energy savings, this can also improve transparency by reducing condensation on the warm air side. The hermetically sealed space is filled with dehydrated air or gas. The panes are connected by a spacer, using sealants to reduce water vapour penetration. The whole unit is hermetically assembled by a secondary edge seal (polysulfidpolymer or silicone) which gives structural robustness to the insulating glass. The spacer contains a desiccant that absorbs humidity from within the air space. The IGU is made manually or by automated machinery.

In combination with special coatings (see Section 1.3.7), modern IGUs achieve overall heat transfer coefficients (U-values) of $1.1\,W/m^2K$ for double-glazed units and $0.7\,W/m^2K$ for triple-glazed units. All types of annealed, heat strengthened or fully tempered monolithic or laminated glasses can be used in IGUs. The space between the glasses may contain interior muntins.

Figure 1.16:
Double-glazed insulating glass unit, principle build-up.

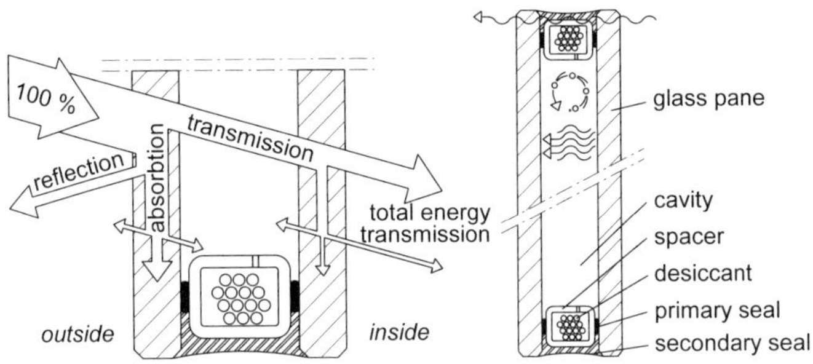

1.3.5 Curved glass

Curved glass, formerly known as 'bent glass', is glass that has been heated past its softening point and formed into a curved shape, usually by draping the softened glass over or into a mould. A mould release agent prevents direct contact between the mould and the glass. While curved glass is commonly used for automotive glazing, it is not often found in architectural applications. The main reasons are the high manufacturing costs and the tolerance related difficulties encountered with the production of curved insulating or laminated glass units.

Glass may be curved along one or both axes. Uniaxial curving is generally achieved by sag bending which simply allows the heated glass to take on the form of the mould by its own weight. For doubly curved shapes, the glass must be pressed into the mould. Using special tempering equipment with individually adjustable rollers, curved glass can be thermally tempered as long as the radius is not too small and if the bending angle does not exceed 90 degrees. If small radii or larger bending angles are required, chemical tempering may be an alternative.

A geometric method proposed by *Schober* transforms the curved surfaces into a planar quadrangular mesh thus avoiding the need for expensive curved glass in the construction of complex free-form shells. The method is based on the translation of one spacial curve against another [299].

1.3.6 Decorative surface modification processes

The following are the most common modification processes used to obtain decorative effects:

- *Acid etching* is a process where the glass surface is treated with hydrofluoric acid. Acid-etched glass has a distinctive, uniformly smooth and satin-like appearance. *Sandblasting* produces a similar effect, but with a rougher texture. Glass treated with one of these processes, also referred to as *frosted glass*, is translucent, obscuring the view while allowing light transmission. Acid-etched and sand-blasted patterns are very durable and not subject to degradation due to weathering.
- To produce *enamelled or screen printed glass*, a ceramic frit colour, consisting of glass powder (70–95%) and pigments (5–30%), is sprayed onto the cooled annealed glass and then burned into the surface during the tempering process. The surface may be covered totally or partially. Any pattern or image can be obtained by spraying the colour through a screen. Enamel coatings have a thickness of about 10–100 μm and are usually applied to the gas side of float glass. The colour does not prevent the production of laminated glass using PVB or resin, but it reduces the mean value of the bending strength by about 25–40%. The scatter of the strength is reduced, too. Dark coatings are somewhat problematic because they may trigger thermal breakage. Ceramic coatings should not be applied to surfaces exposed to weathering in order to avoid degradation.
- *Ink-jet printing* on glass surfaces is possible today, using special colours. No data for the fastness to light is available yet, however the durability is expected to be inferior to that of enamelled glass
- *Body-tinted glass* is produced by adding metal oxides (iron oxide, cobalt oxide, titanium oxide and others) to the constituent materials during the production of float glass. These metal oxides produce a consistent colour throughout the glass thickness. Various bluish, greenish, brownish, greyish and reddish tones are available. As the colour is very sensitive even to little changes of the glass composition, an exact colour match between different production lots is difficult to obtain.
- *Patterned glass* is glass with an embossed pattern on one or both surfaces. It is mostly produced using the cast process (see Section 1.1.2) by means of patterned rollers. The strength of patterned glass is usually much lower compared to flat glass.
- *Abrasion* is a method of shallow, decoration grinding using a diamond wheel.

Figure 1.17: Examples of decorative surface modification processes: patterned glass (left), ceramic frit (middle), acid etched pattern (right).

1.3.7 Functional coatings

Coating processes

Hard coatings Hard coatings are commonly applied using a chemical vapour deposition process. In this process, also known as pyrolytic coating, a gaseous chemical mixture is brought in contact with the hot glass substrate (600–650 °C) and a pyrolytic reaction occurs at the surface of the substrate leading to the deposition of a coating which bonds to the glass. Because of the high temperatures required, the coating process is integrated in the float process or the annealing lehr, which is why it is also called on-line coating. A variety of materials ranging from pure metals and oxides to mixed oxide/nitrides can be commercially deposited. An alternative method of applying hard coatings is dip coating. In this process, the glass is dipped into the coating solution and then heated up to 650 °C.

Pyrolytic coatings are very hard. They are scratch resistant, temperable and bendable and can even be applied to exterior faces of glass lites. On the other hand, they are not as flexible as off-line coatings. Only a maximum number of two layers can be applied at once. An example of a popular pyrolytic coating is reflective glass [176, 278].

Soft coatings Soft coatings can be applied to the glass surface by various processes such as dip coating, chemical or physical vapour deposition. The predominant soft coating technique is Magnetron sputtering in which sputtering is performed in a vacuum process by applying a high voltage across a low-pressure gas (usually argon) to create a plasma of electrons and gas ions in a high-energy state. During sputtering, energized plasma ions strike a target, composed of the desired coating material, and cause atoms from that target to be ejected with enough energy to travel to, and bond with, the glass surface. By the use of a planar magnetron, the plasma is confined to the region closest to the target plate, which vastly improves the deposition rate. The coating is carried out in several vacuum chambers with different targets.

Magnetron sputtering allows for the production of high performance, multi-layer coatings using different materials. The process is very precise, flexible and gives very constant coating quality. It makes it even possible to exactly reproduce some specific coating after many years.

The disadvantage of soft coatings is their susceptibility to aggressive environments (e. g. polluted air) and mechanical damage. This makes it necessary to protect soft coatings with a protective layer or assemble them on the cavity oriented surfaces of double-glazed units. A popular application of soft coatings is in the manufacture of low-emissivity glass. [9, 176, 278]

Common coatings

Solar radiation that reaches the earth's surface consists of about 3% short-wave ultraviolet (UV) radiation, 42% visible light (wavelengths from about 380 nm to 780 nm) and 55% long-wave infrared radiation (IR). Most energy is contained in the invisible IR. The strategy for solar protection is, therefore, to block as much IR as possible without reducing the transmittance in the visible spectrum. *Solar control coatings* achieve this by a combination of absorbtion and reflection.

Low-emissivity (low-e) coatings are sputtered or pyrolytic, transparent metallic or metallic oxidic coatings that save energy and increase comfort inside a building by reducing heat loss towards the environment. This heat loss affects both energy consumption and the comfort levels of people working close to glazed surfaces. Low-e coatings are predominantly transparent for visible light, but reflective in the long-wave infrared range and able to reduce the emissivity of glass (see Section 1.2.2) from 0.84 to about 0.05. They are soft coatings and are normally used in IGUs (cf. Section 1.3.4) and applied to the cavity surface of the innermost glass pane.

There is a vast choice of coatings for various purposes available in the market. Combining several properties, e. g. low-e and solar control, within a single coating becomes increasingly popular. Manufacturers are always eager to provide up-to-date information.

1.3.8 Switchable glazing

The extensive use of large area glazing particularly in façades poses major challenges in terms of user comfort and the conservation of energy in buildings. This challenge is expected to increase further as building regulations become more stringent in terms of energy conservation in an attempt to reduce carbon emissions.

Glazed façades are often required to meet transient and often conflicting performance requirements such as the need to mitigate energy loss, unwanted energy gain and visual discomfort from glare as well as to provide the desirable levels of visual transparency. One approach is to provide a smart and truly responsive façade where the properties of the glass change to actively control solar gain, daylight and glare. The emerging technologies of 'smart glass' or 'chromogenic switchable glazing' offer variable thermal and light transmittance characteristics by responding dynamically to external references such as temperature and light. Such products have the potential to control the amount of visible radiation and IR that enters the building and thus optimize energy efficiency and comfort levels for any given external climatic condition.

The operation of chromogenic switchable glazing is based on the incorporation of materials or devices that allow the optical properties of the glass to change in function of an external stimulus. A change in the reflectance, absorptance or scattering manifests itself in a colour change. It can affect only a part or the whole range of radiation in the solar spectrum, and it can occur passively or actively.

Passive or 'self-adjusting' chromogenics are environmentally driven systems that directly respond to changes in ambient light conditions or temperature and include the photochromic, thermochromic and thermotropic materials. *Active* or 'externally activated' systems require an external electrical current to drive the change in properties and include the electrochromic, liquid crystal, suspended particle and gasochromic devices. The fundamental difference between these two types of chromogenic glazing is that self-adjusting systems are not linked to any external devices whereas externally activated systems are regulated through a transducer that may be controlled by the user or by a set of sensors that is linked to the building management system. More detailed information on the range of chromogenic glazing available is found in [6, 71, 266, 346], however a brief overview of the specific systems is provided below.

Self-adjusting systems

Photochromic glazing Photochromic glass reduces light transmittance by darkening when exposed to ultraviolet radiation. This darkening phenomenon occurs due to the chemical composition of the glass itself that includes photosensitive silver halide crystals. The energy delivered by wavelengths between 300 and 400 nm break down the crystals, therefore causing increased absorption of the visible wavelengths and thus darkening of the glass. This process is reversed when the source of ultraviolet radiation is removed [345]. Photochromic glass is durable and has a long service life. The visible radiation transmission ranges from about 85% to about 25% in the two states, however, the complexity of the manufacturing process, the high cost of its components and the rather slow reaction times have limited its production to small non-architectural quantities and sizes (e. g. photochromic eyeglasses).

Thermochromic glazing Thermochromic glass alters its optical properties in response to changes in temperature. This is caused by a thin layer of thermochromic material that is applied on the glass surface. When the temperature of the thermochromic material rises to a set temperature, a reversible chemical reaction (phase transformation) is induced that causes a change in the material's transmission properties. Transition metal oxides such as vanadium dioxide (VO_2), for example, change from a semiconductor state with low absorption in the infrared range to a metallic state exhibiting infrared reflectivity when they absorb a certain amount of heat energy [71]. In the metallic state the thermochromic layer operates as a low emissivity coating. Thermochromic glass can thus control both transmittance and infrared emissivity of a glazed façade.

Issues that still need to be addressed before the commercialization of thermochromic glass is made possible include durability, low light transmittance, setting of the transition temperature and the yellow colouration of the darkened state.

Thermotropic glazing Thermotropic materials respond to changes in temperature by altering their optical properties, similar to thermochromics. However, a difference in the internal mechanism of the property change gives thermotropics the potential to go through a radical transformation from a clear, light-transmitting semiconductor state to an opaque, light-scattering insulator state. When thermotropic materials are heated, both their reflective properties and their thermal conductivity are altered. Thermotropics are the only chromogenic materials to date that are able to control heat transfer not only through radiation but also through conduction [6]. However, they do so at the expense of transparency and view. The principle of the operation of thermotropic materials is the combination of at least two materials with different refractive indices such as water and a polymer (hydrogel), or two different polymers (polymer blend). In its original state, the mixture is homogeneous. As the temperature rises, the molecular structure of the polymers changes from stretched chains to clumps that diffuse light, such that most solar radiation is reflected [284]. For a typical thermotropic layer, the solar energy transmission ranges from 80–90% to between 10 and 50%, depending on the composition of the specific material. Light transmission values follow a similar range.

Several technical problems with hydrogels, such as inhomogeneity during switching, UV stability, cycle lifetime and the requirement for tight edge seals, have complicated the development of thermotropic glazing units. A low-e glazing unit that incorporates a thermotropic film and a layer of transparent insulation is at present available, but the manufacturer warns that visual changes or changes with regards to switching behaviour may occur over its lifetime.

Externally activated systems

Liquid crystal glazing Liquid crystal (LC) technology has already been used in buildings and several liquid crystal glass products are available. LC glazing is a laminated glass comprising two sheets of glass and a liquid crystal film. The LC film consists of two outer layers of polyester that are coated with a transparent conductor and of a polymer matrix that contains the liquid crystals. When no voltage is applied, the liquid crystal molecule chains are randomly scattered and the LC system is translucent opal white. When a voltage is applied, the molecules align with the lines of the electric field and the film appears almost transparent. Open circuit memory is not possible, i. e. the device remains transparent only for as long as the electric field is maintained.

Large LC panels of up to 1000 mm × 3000 mm have already been produced. Switching between the clear and diffuse states is literally instantaneous. However, LC panels cannot control the light and heat flow through the glazing. They do not actually exhibit variable transmission characteristics since they only affect the way light is transferred and not the quantity of radiation that is allowed to pass through. Furthermore their high production cost, their instability when exposed to ultraviolet radiation and the obstruction of view in the obscure state explain why their use in architecture is usually restricted to internal applications, such as privacy partitions.

Suspended particle glazing Suspended particle devices (SPDs) are similar in character to liquid crystal devices. They incorporate an active layer that contains needle-shaped dipole particles that are uniformly distributed in an organic fluid or film. The active layer is laminated or filled between two transparent conductors on polyester. In the 'off' condition, the particles are randomly orientated and absorb a large part of incident radiation. When a voltage is applied, the particles align with the electric field and radiation transmission is increased. The device changes from a coloured state, when it appears dark blue, to a clear state; the degree of the tint can be varied depending on how much current is applied and the change is almost instant. An SPD does not scatter light when it is in the darkened state and thus view is not obstructed at any stage of colouration. Suspended particle panels up to 1000 mm × 2800 mm for architectural applications are at present commercially available. Light transmission values for such panels range from about 0.5–12% in the dark state to 22–57% in the clear state. Shading coefficient values range from 47–57% to 64–80% respectively. This means that although visible radiation can be remarkably reduced by darkening the device, the shading coefficient values remain relatively high. The heat gains thus remain considerable even in the dark state. Therefore, the light to heat gain ratio cannot be considered favourable for solar radiation control.

Electrochromic glazing Electrochromic glazing is the most popular and most complex of all switching glazing technologies. Various electrochromic devices have so far been developed; the ones intended for architectural applications incorporate solid electrochromic films and they consist of a thin multi-layer assembly that is typically sandwiched between two panes of glass. They rely on the colouration of solid anodic or cathodic electrochromic films to modulate their optical properties. Anodic films colour upon electrochemical oxidation whereas cathodic films rely on electrochemical reduction for colouration. These reactions involve the transfer of ions into and out of the electrochromic films and thus, electrochromic devices require a component where ions can be stored when removed from the electrochromic film. This requirement is usually met either by incorporating an ion storage layer or by coupling an anodic and a cathodic electrochromic film.

The most widely used cathodic electrochromic film consists of tungsten oxide (WO_3) because it has the greatest variation between the clear and the dark states. Electrochromic devices remain specular at all stages of colouration and blue colour is the most common result of the darkening process. The visible radiation transmission of typical electrochromic devices ranges from 70–50% in the clear state to 25–1% in the fully coloured state. The shading coefficient values range from 67–60% to 30–1%. As the electrochromic device colours, transmission is kept at higher levels in the visible part of the solar spectrum than in the infrared part, resulting in a high light to heat gain ratio. The voltage required for the operation is small and it only needs to be applied during switching [235]. The switching times depend on the type of the device and the size of the window; typically full colouration is achieved in 5 to 10 minutes.

Common problems faced in the quest for a reliable, large-scale electrochromic device are long term degradation, sensitivity to environmental conditions and the relatively long switching times which rise with increasing device size. These issues have been addressed and partially solved and at present there are a few electrochromic glazing products for architectural applications available in the market.

Gasochromic glazing Gasochromic systems produce a similar effect to electrochromic systems. Their operation is based on the principle that thin films of tungsten oxide colour in the presence of hydrogen gas. Gasochromic devices consist of two panes of glass, which are coated with a layer of tungsten oxide and a catalyst respectively. When diluted hydrogen is introduced in the cavity between the two glass panes, the tungsten oxide reacts with hydrogen and colours. To return to its original transparent state, the cavity is purged with another gas, usually oxygen. The desired mix of hydrogen and oxygen is diffused in the cavity by a pump connected to a small electrolysis unit that decomposes water. The gas circulates in a closed cycle and is reconstituted as water, in the presence of a catalyst, when the pump is switched off [174]. Visible transmittance of 75% to 18% and total solar energy transmission of 74% to 14% have been obtained [235].

The main advantages of gasochromic devices are their simple coating structure, the high transmission levels in the clear state and the short switching times. The main technical difficulties in the construction lie on the gas injection system, the plumbing of the gas tubes and the avoidance of water build-up when hydrogen atoms are added [71]. Gasochromic glazing is not commercially available at the moment.

Most of the chromogenic glazing systems described above are currently being researched and developed. It is therefore difficult to determine the best system at this stage. *Table 1.8* provides a brief overview of the main advantages and disadvantages of these systems. An important distinguishing factor is that between self-adjusting (passive) and externally activated (active) systems. Although the idea of incorporating a self-adjustable light filter in glazed façades may appear attractive, the lack of external control may compromise the performance of environmentally driven systems in two main ways. Firstly, in order to achieve optimum performance of a glazed façade, the proportion of light, heat and view provided by it must be able to change and adapt to varying and often conflicting requirements. The optimization of only one of these three factors is unlikely to result in the ideal response for the other two factors throughout the year. Secondly, at least a certain degree of local user control on the system is preferred as this has a significant effect on comfort which is in turn the major influence on productivity and the economics of commercial buildings.

Table 1.8: Comparison of switchable glazing types.

Type	Advantages	Disadvantages
	Self adjusting systems	
Photochromic	Long life	High cost, small panels
Thermochromic	Low emissivity	Poor durability, low light transmittance, yellowish colour
Thermotropic	Excellent thermal performance	Degrades on exposure to ultra violet radiation
	Externally activated systems	
Liquid crystal	Established technology	Very poor thermal performance
Suspended particle	Established technology	High heat to light transmittance
Electrochromic	High light to heat transmittance, relatively low cost	Slow switching times for large panels, insufficient durability
Gasochromic	Very rapid switching times	Complex due to gas injection

1.3.9 Other recent glasses

Self-cleaning glass

Self-cleaning glass is made by applying a microscopically thin (approximately 40 nm thick) titanium oxide (TiO_2) based coating onto float glass by chemical vapour deposition (see Section 1.3.7).

The titanium oxide based coating has both semiconductor and hydrophilic properties. It therefore performs two functions: firstly, it absorbs UV light to promote oxidation and reduction of organic materials and to reduce the adherence of inorganic dirt; secondly, it reduces the contact angle of water with glass and thus induces the raindrops to be dispersed over a wide surface, rather than forming droplets, and run off in a 'sheet' to wash the loosened dirt away.

For further details of the physical and chemical characteristics of self-cleaning glass, readers may refer to [294].

Embedded LEDs

Light emitting diodes (LEDs) may be embedded into a laminated glass unit by using a 2 mm thick cold poured interlayer. The power supply to the LEDs is provided by a standard low voltage supply via a virtually invisible conductor plate on the internal surface of one of the glass plates. Standard float glass sizes may be used and the LEDs may also provide special effects such as flashing indicators and running lights.

Dichroic glass

Dichroic glass changes colour in different environments. Colours vary depending on the intensity of natural light, the angle of view and the background lighting conditions. This effect is achieved by the application of selective metal oxide coatings in a thickness of less than 100 nm to a range of base glasses. Light is reflected from the junction of these layers to different intensities, the reflection increasing as the refractive indices of the layers are further apart. By selecting the number, sequence, thickness and optical properties of the layers, certain wavelengths reflect strongly and others are transmitted through the glass.

Photovoltaic glass (PV)

The direct conversion of light into electricity known as the photovoltaic effect was discovered in 1839 by Edmund Becquerel. However the real breakthrough in solar research did not come until the 1960s with the development of solar sails used in space travel. The first mass produced applications began with small solar cells used in solar-powered pocket calculators. The recent emphasis on renewable energy sources and the simultaneous industrial production of efficient PVs have provided a further boost for building integrated PVs. Several governments provide subsidies for the use of PVs and typical pay back periods are currently between 3 and 7 years.

PV consists of laminated glass with integrated solar cells to convert solar energy into electricity. The solar cells are embedded between two glass panes by means of an EVA (ethylene vinyl acetate copolymer) interlayer. The EVA interlayer is preferred to the traditional PVB used in standard laminated glass (cf. Section 1.3.3) as the former does not require autoclaving, which would damage the solar cells. Each individual cell has two electrical connections, which are linked to other cells in the module, to form a system which generates a direct electrical current.

There are a wide range of solar cells available, though the bulk of the material in use today is semi-conductor grade silicon. The PVs embedded in glass are generally known as thick crystalline silicone cells which are between 200 and 300 microns thick. Current commercial modules achieve around 15% efficiency whereas research cells are at 24% efficiency.

Various cell sizes are produced by different manufacturers and spacing between the cells can be varied in each direction, thus allowing a degree of transparency through the PV panel. The front pane of glass is generally a heat strengthened low iron glass. The inner pane of glass can be of any type and may include a low-e coating to improve thermal performance. PV panels may form part of insulating glass units (cf. Section 1.3.4) and panel sizes in excess of 3000 mm × 2000 mm are available.

1.3.10 Relevant standards

Table 1.1 gives an overview of important standards for processed glass products. For standards on basic glass products, see *Table 1.9*.

Table 1.9: Important standards for processed glass products (shortened titles).

Standard	Title
EN 1863-1:2000 [133]	Heat strengthened soda lime silicate glass – Part 1: Definition and description
EN 1863-2:2004 [134]	Heat strengthened soda lime silicate glass – Part 2: Evaluation of conformity/Product standard
EN 12150-1:2000 [101]	Thermally toughened soda lime silicate safety glass – Part 1: Definition and description
EN 12150-2:2004 [102]	Thermally toughened soda lime silicate safety glass – Part 2: Evaluation of conformity/Product standard
EN 14179-1:2005 [123]	Heat soaked thermally toughened soda lime silicate safety glass – Part 1: Definition and description
EN 14179-2:2005 [124]	Heat soaked thermally toughened soda lime silicate safety glass – Part 2: Evaluation of conformity/Product standard
EN 13024-1:2002 [117]	Thermally toughened borosilicate safety glass – Part 1: Definition and description
EN 13024-2:2004 [118]	Thermally toughened borosilicate safety glass – Part 2: Evaluation of conformity/Product standard
EN 14321-1:2005 [125]	Thermally toughened alkaline earth silicate safety glass – Part 1: Definition and description
EN 14321-2:2005 [126]	Thermally toughened alkaline earth silicate safety glass – Part 2: Evaluation of conformity/Product standard
EN 12337-1:2000 [103]	Chemically strengthened soda lime silicate glass – Part 1: Definition and description
EN 12337-2:2004 [104]	Chemically strengthened soda lime silicate glass – Part 2: Evaluation of conformity/Product standard
EN 1096-1:1998 [97]	Coated glass – Part 1: Definitions and classification
EN 1096-2:2001 [98]	Coated glass – Part 2: Requirements and test methods for class A, B and S coatings
EN 1096-3:2001 [99]	Coated glass – Part 3: Requirements and test methods for class C and D coatings
EN 1096-4:2004 [100]	Coated glass – Part 4: Evaluation of conformity/Product standard
ISO 12543-1:1998 [205]	Laminated glass and laminated safety glass – Part 1: Definitions and description of component parts
ISO 12543-2:2004 [206]	Laminated glass and laminated safety glass – Part 2: Laminated safety glass
ISO 12543-3:1998 [207]	Laminated glass and laminated safety glass – Part 3: Laminated glass
ISO 12543-4:1998 [208]	Laminated glass and laminated safety glass – Part 4: Test methods for durability
ISO 12543-5:1998 [209]	Laminated glass and laminated safety glass – Part 5: Dimensions and edge finishing
ISO 12543-6:1998 [210]	Laminated glass and laminated safety glass – Part 6: Appearance
EN 14449:2005 [127]	Laminated glass and laminated safety glass – Evaluation of conformity/Product standard
EN 1279-1:2004 [107]	Insulating glass units – Part 1: Generalities, dimensional tolerances and rules for the system description
EN 1279-2:2002 [108]	Insulating glass units – Part 2: Long term test method and requirements for moisture penetration
EN 1279-3:2002 [109]	Insulating glass units – Part 3: Long term test method and requirements for gas leakage rate and for gas concentration tolerances
EN 1279-4:2002 [110]	Insulating glass units – Part 4: Methods of test for the physical attributes of edge seals
EN 1279-5:2005 [111]	Insulating glass units – Part 5: Evaluation of conformity
EN 1279-6:2002 [112]	Insulating glass units – Part 6: Factory production control and periodic tests
ASTM C 1048-04 [12]	Standard Specification for Heat-Treated Flat Glass – Kind HS, Kind FT Coated and Uncoated
ASTM C 1172-03 [13]	Standard Specification for Laminated Architectural Flat Glass
ASTM C 1376-03 [14]	Standard Specification for Pyrolytic and Vacuum Deposition Coatings on Flat Glass
ASTM C 1422-99 [16]	Standard Specification for Chemically Strengthened Flat Glass
ASTM C 1464-06 [17]	Standard Specification for Bent Glass
ASTM C 1503-01 [18]	Standard Specification for Silvered Flat Glass Mirror

Chapter

2 General Design Guidelines

2.1 The design process

This text has been compiled in collaboration with the following experts:
Christoph HAAS

2.1.1 Particularities of glass structures

The overall design procedure for structural glass elements is not unlike other structural materials i. e. it is essentially an iterative process that relies on a combination of rules of thumb, more accurate analytical methods and prototype testing. The use of these three techniques varies throughout the design process. Quick, approximate methods are primarily used at early design stage to test alternative schemes and at a later stage for verifying the more accurate calculations; more accurate methods are employed during detailed design stages; prototype testing is used to verify the design prior to construction.

As with any structure, the designer should establish the fundamental performance requirements before starting any calculations. These requirements include the ultimate limit state that ensures adequate strength to withstand the anticipated actions, namely, material strength, overall structural stability (i. e. the structure is not a mechanism) and elastic stability (i. e. no flexural or lateral torsional buckling). Additional ultimate limit state performance requirements that are particularly relevant to glass deal with fail-safe concepts, ranging from criteria for overall structural robustness to requirements for the post-breakage structural behaviour of individual glass elements. Serviceability limit state requirements normally include limiting deflections and/or vibrations, movement tolerances and aesthetic criteria. It is understood that all the ultimate and serviceability limit states should be satisfied in order to ensure structural adequacy.

The standard elastic design method used with most construction materials is known as the maximum stress approach. In this approach the engineer sizes a structural element by ensuring that the maximum stresses caused by an action do not exceed the strength of a material at any position on that element. Most engineers therefore implement structural design from a few fundamental constants, the strength of the material being one of them.

However, the strength of glass depends on a number of factors (cf. Chapters 1 and 3). This explains the lack of a single accurate value for the design strength of glass and why the maximum stress approach is unsuited for designing structural glass elements. Furthermore, glass shows an almost perfectly elastic, isotropic behaviour and exhibits brittle fracture. This inability to yield plastically means that glass cannot redistribute local stress concentrations by local yielding. Structural glass elements are, therefore, extremely susceptible to stress concentrations and failure occurs without warning. This 'unforgiving' brittle nature is crucial in the design of glass elements and connections, requires a greater attention to detailing and much tighter fabrication/construction tolerances than connections in steel or timber structures. In order to avoid unexpected stress concentrations, the design model must account for all relevant aspects and be analysed thoroughly. A good structural model of a glass structure should account for conventional actions due to load, temperature differences, imposed deformations and constraints, as well as the detailed geometry, the stiffness of all components including support bracketry and fixings as well as fabrication/installation tolerances including out of plane distortions and closeness of fit.

Consequently, in order to undertake structural glass design the engineer must have an in-depth knowledge of the specific properties of glass (and the other materials employed in the glass structure), select the appropriate design method that faithfully models the glass structure in question and carry out sensible detailing.

2.1.2 Risk analysis

Safety considerations are at the interface between the intended use of a structure and its design and sizing. Although it is impossible to provide sufficient resistance for every conceivable threat, any structure should perform satisfactorily under foreseeable circumstances. A set of design situations (sometimes referred to as design cases) should be established from relevant hazard scenarios and corresponding service situations. Design and sizing is then based on these design situations and on the assumption that the structure will be built as planned. Appropriate safety factors have to be taken into account.

The resistance of a glass element is very sensitive to the flaws on its surface. In addition to standard hazard scenarios including loads and constraint stresses, surface damage hazard scenarios should, therefore, be considered for design. Such hazard scenarios represent factors that cause severe surface damage without instantaneous failure such as accidental impact, vandalism, or heavy wind-borne debris.

A structure can be in danger for two fundamentally different reasons: It may be subjected to actions that it was not designed for or the structure may not have been planned or built properly. Both threats can be countered in a number of ways (*Figure 2.1*):

- The structure can be designed to withstand the threat.
- Measures to reduce the likelihood of the threat can be taken.
- The threat can be accepted as an unmitigated risk.

If the structure is to be designed to withstand particular threats, corresponding design situations must be defined. Typically, these are accidental design situations that may require lower safety factors than non-accidental design situations and can be considered in parallel with the latter for the sizing of the structure. This approach is generally unsuitable for most glass structures as it would entail a 'no-break' scenario that often results in very

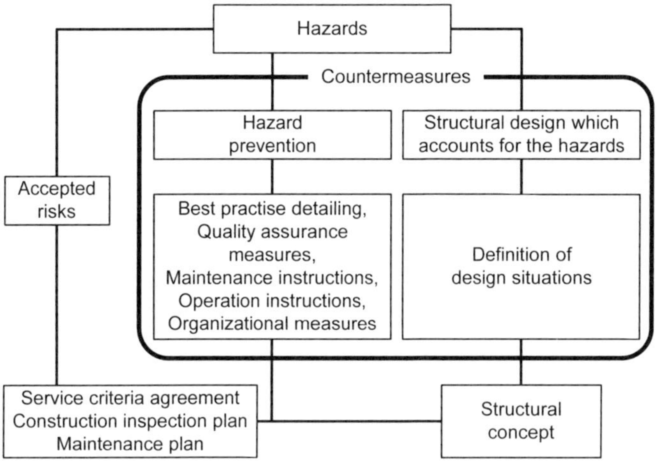

Figure 2.1: Hazards and countermeasures.

thick glass elements and visually obtrusive sub-frames and connections. Alternatively, measures to avoid a threat may involve:

- design modifications (e. g. improved redundancy and alternative load paths),
- proper quality assurance during planning and construction stages,
- proper maintenance,
- adjustments in the way the structure will be used,
- permanent additional safety measures (e. g. protection against car impact),
- temporary additional safety measures during certain service situations (e. g. protection of glass edges during delivery of bulky goods).

From the viewpoint of reliability, it is preferable to mitigate a risk by incorporating the appropriate measures in the first instance rather than relying on measures that must be taken during the entire lifetime of a structure. This is because measures can be implemented more reliably during the design and construction phases rather than being enforced during the entire service life of a structure.

The third possible approach is to accept a threat as an unmitigated risk. This may be appropriate if a risk is deemed sufficiently improbable to occur, if its consequences are considered sufficiently small or if a combination thereof justifies such a decision. Systematic approaches to assess whether a risk is acceptable or not are e. g. given in EN 1991-1-7:2006 [137]. Acceptable levels of risk are generally high if the person at risk can influence the risk and is taking it voluntarily (e. g. a mountain climber). The opposite is typically true for building structures: Acceptable risk levels are very low because the people at risk have little or no influence on the risk and are not even aware of taking a risk. Furthermore, the potential consequences of a structural failure are often very severe in terms of the number of affected people and potential economic losses.

Clearly, the intended use, maintenance procedures, permanent and temporary safety measures as well as accepted risks must be discussed in detail with the owner and if possible also with the users of the structure. All decisions must be documented and the key issues must be made available to all involved parties by including them in the operations and maintenance manuals.

2.1.3 Post-breakage behaviour and robustness

Post-breakage structural capacity can be made available either at the level of an individual structural element or at the level of the structural system. In the former case, an element should fail only partially or at least in a ductile manner. In the latter case, brittle failure of an individual element may occur, but the structure must be able to redistribute loads to other elements, thereby providing redundancy. While redundancy may prevent the failure of large parts of a structure, the failing element itself may pose a considerable local threat (e. g. overhead glazing). Glass does not possess any inherent ductility and disintegrates after failure. Any remaining load-bearing capacity must, therefore, be achieved by additional means. Laminated glass with a polyvinyl butyral (PVB) interlayer is often able to provide an adequate level of post-breakage performance by interaction of the PVB film and the glass fragments (cf. Section 1.3.3). Contrary to popular belief, however, laminated glass units with a PVB interlayer do not always guarantee post-breakage stability. In the case of laminated glass composed of fully tempered glass plates, the highly fragmented panels are often unable to mobilize an arching or locking action that is essential for a degree of post-breakage stability, particularly if the broken tempered glass is subjected to high levels of compression. Consequently such laminated glass panels will normally sag like a wet towel (*Figure 2.2*). In these cases, post-breakage structural capacity relies solely on the tensile strength of the PVB interlayer, which has a tendency to tear. Furthermore, the panel often slides from its supports or out of clamps as a consequence of the large deformation.

Three stages of flexural behaviour in laminated glass are shown in *Figure 2.3*: Stage 1, where both sheets of glass are intact; Stage 2, where the bottom sheet has fractured and the top sheet is carrying all the loads; and Stage 3 where the top sheet has also fractured but the fragments in the top sheet lock together in compression and combine with a tensile stress in the interlayer to provide some further post-breakage resistance. The extent of the flexural post-breakage resistance provided by Stage 3 depends on the stiffness and tensile strength of the interlayer and on the type of glass used in the top layer of glass. If the laminated glass is composed of annealed or heat strengthened glass, which break into large shards, the PVB interlayer is able to hold the glass fragments in place, thus enabling compressive forces to be transmitted through the broken glass and providing a limited amount of post-breakage stability. As mentioned in the preceding paragraph, this is much less the case for fully tempered glass because of the small size of the fragments.

A similar post-breakage performance can be achieved by using a combination of annealed or heat strengthened glass panes with fully tempered panes, as long as the fully

Figure 2.2: Fully tempered monolithic (left) and laminated (right) glass after breakage [272].

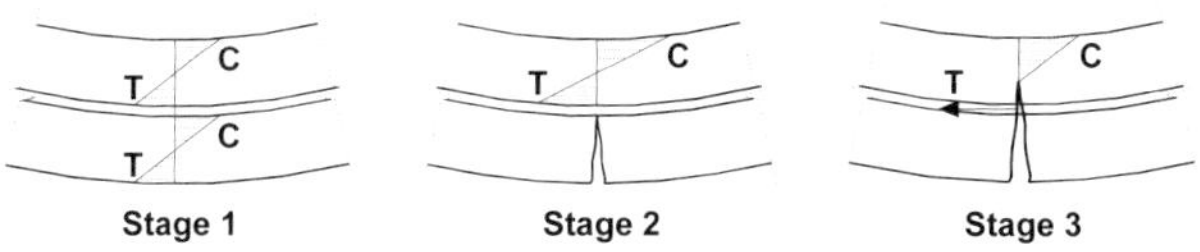

Figure 2.3: Three stages of flexural behaviour in laminated glass showing the post-breakage stress distribution [232].

tempered glass panes are located on the tension side of the laminated unit. An alternative way of increasing the post-breakage structural capacity of an individual element is the use of steel or carbon fibre elements to act compositely with the glass elements. An adequate connection between such elements and the glass is crucial to achieve the desired post-breakage performance [268, 270].

If post-breakage structural capacity is to be achieved by load redistribution, the prevention of progressive failure is a key factor. Alternative load paths must be in place, which means that elements neighbouring a failed element must be able to withstand the additional loads caused by load redistribution. If this is not properly accounted for in their design, progressive failure leading to a partial or even total collapse of the structure is unavoidable.

It is important to think in terms of actual threats. Some threats, e. g. an impact, may be singular, local events that affect only one particular structural element at a random point in time. Such threats are improbable to coincide with the maximum intensities of other actions like wind or snow. Although load redistribution will increase the load on neighbouring elements, this load may still be well below their maximum design loads. This assumption does not hold true in the case of a threat that is correlated with other actions. The threat of falling trees, for instance, is correlated with strong winds. The load redistribution requirements in such cases are more onerous and the resistance of all elements involved may need to be increased in order to prevent progressive failure.

For threats that affect large parts of the glass structure, e. g. explosions, load redistribution may not be feasible. The simultaneous failure of multiple structural elements can lead to spans that are impossible to bridge without unreasonable aesthetic, technical or economic consequences. In such cases, the design approach is often to mitigate the risk by providing protective glazing design that involves measures to reduce injuries from broken glass. This approach is similar in nature to glazing design for blast loading (cf. Section 2.2.6).

Project specific post-breakage tests are often the only way to ensure sufficient post-breakage performance (see Section 8.1).

2.2 Actions on glass structures

This text has been compiled in collaboration with the following experts:
Dr. Frank WELLERSHOFF

2.2.1 Particularities of glass structures

The actions on glass structures are largely similar to the actions on most other building structures and include self weight, dead loads, live loads, wind loads, snow loads, thermal stresses, pressure differences, impact loads, blast loads and seismic loads. However the resistance of glass elements is very sensitive to flaws on their surface, therefore damage caused by accidental impact, vandalism, heavy wind-borne debris and the like may also be important actions to be considered for design (see Section 2.1.2).

Another particularity of glass is that the entire stress history (caused by load fluctuations, load durations etc.) has a major influence on sub-critical crack growth and therefore on the inherent strength (cf. Section 1.2.2). Consequently, complete action history models are required in order to design glass elements in an accurate manner. This is a fundamental difference from other materials such as steel or concrete, for which only extreme values or extreme value distributions of actions are normally used for design (except for fatigue considerations). However, most of the current glass-related codes, specifications and guidelines provide information on extreme values only. If the inherent strength of glass is neglected and only the residual surface stress is considered for design (e. g. in the case of HSG or FTG) extreme value models are sufficiently accurate and the stress history may be ignored (cf. Chapter 6).

Design action intensities, reference time periods, partial safety factors and the like vary across countries and standards. Since such quantities are inter-related, it is essential that only compatible data is used. Characteristic action values from European standards, for instance, cannot be combined with partial safety factors of some other standard and vice versa.

The main European standard on actions is EN 1991-1-7:2006 [137]. It covers standard actions such as dead loads, live loads, wind loads and snow loads. The self weight of glass is given in EN 572-1:2004 [148] (see *Table 1.5*). Partial factors as well as guidelines on the design cases are generally given in the material-specific standards or guidelines (see Chapter 4).

Section 2.2.2 to Section 2.2.9 below provide additional information related to glass-specific actions that are only partially covered in current standards.

2.2.2 Wind loads

Wind induced pressures give rise to dynamic actions that are normally described by two simplified parameters, the mean wind speed and the turbulence intensity. These parameters are normally used to define an equivalent static wind load on glazing. Such an approach is acceptable in most design situations since the natural frequency of a façade is normally significantly higher than the periodic occurrence of localized gusts. However, load amplification can occur in slender, large span façades such as suspended glazing and cable stayed structures, where the natural frequency of the local structure may be below 1 Hz.

An accurate mathematical prediction of the lifetime of a glass panel subjected to wind loading is still elusive due to the complexities and various parameters involved. In essence, each gust is equivalent to a load cycle on the glass, which may cause subcritical crack growth. This complex load history and the resulting stress history will affect material strength (cf. Section 3.2 and Section 3.3). Furthermore sharp edges and obstructions on glass façades give rise to flow separation and vortices. The statistical variation of the negative pressures that arise from this turbulence do not fit a normal distribution. The effect of this phenomenon on the strength of glass is discussed in [230].

In practice, approximate methods for predicting wind loading on glass façades broadly include the use of national codes of practice and international wind loading standards [137], wind tunnel testing and more recently computational fluid dynamics. A basic review of the accuracy and applicability of each of these different methods is provided in [269]. The designer should be well aware of the limitations of the prediction method used, as unsafe results may be derived if these are not observed. Other design issues related to wind actions on glass include the calculation of internal pressures in buildings, pressure distributions and pressure losses in double façades and load sharing between the individual panes of insulated glazing units (IGUs). Guidance on the first two issues is provided in the wind loading Eurocode EN 1991-1-7:2006 [137] whereas for instance DIN 18008-1:2006 [81], DIN 18008-2:2006 [82], prEN 13474-2:2000 [281] give guidance on the distribution of wind loads in IGUs.

2.2.3 Correlation of wind load and material temperature

PVB-foils in laminated safety glass and adhesives in glued connections are polymers with viscoelastic properties. The mechanical behaviour of these materials depends on the three dimensional spectrum of time, temperature and load. Polymers become weaker with rising temperature and tend to creep under high forces and long load durations. In the event of sustained loads it is therefore common practice to assume that interlayers creep and thereby ignore any shear transfer that occurs from one glass panel to another. However when dealing with short duration loads (e. g. wind loading) it is sensible to account for some shear interaction by superposing the variations in material temperature, load and design lifetime. However, it is unrealistic to assume that the highest wind load coincides with highest temperature and doing so would lead to uneconomic glass thicknesses for resisting wind loads. *Figure 2.4* shows the typical correlation between the gust wind speed and the air temperature in Germany. This figure shows that higher wind speeds occur during storms with air temperatures of 10 to 15 °C which are well below the maximum recorded temperatures that are in the order of 30 °C.

The second effect during storms is that the sky is overcast and the amount of solar energy reaching the glass is substantially reduced. The wind speed and temperature effects may be correlated as shown in *Figure 2.5*. This figure is based on daily measurements of the relative gust wind load and temperature from eight weather stations in Germany for the period between 1970 and 1998. The related material temperature due to absorption

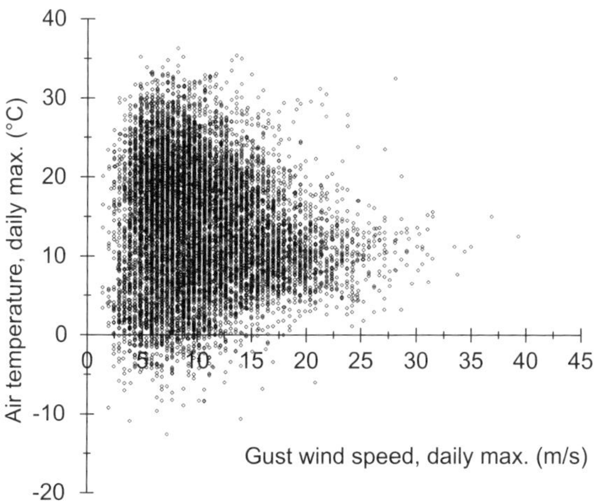

Figure 2.4:
Wind gust speed and air temperature in Aachen, Germany, daily maximum values in 28 years [338].

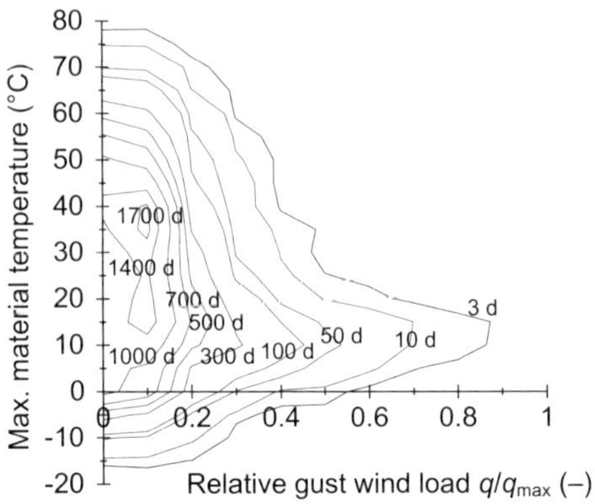

Figure 2.5:
Correlation of wind load and material temperature, Germany, 100 years [338].

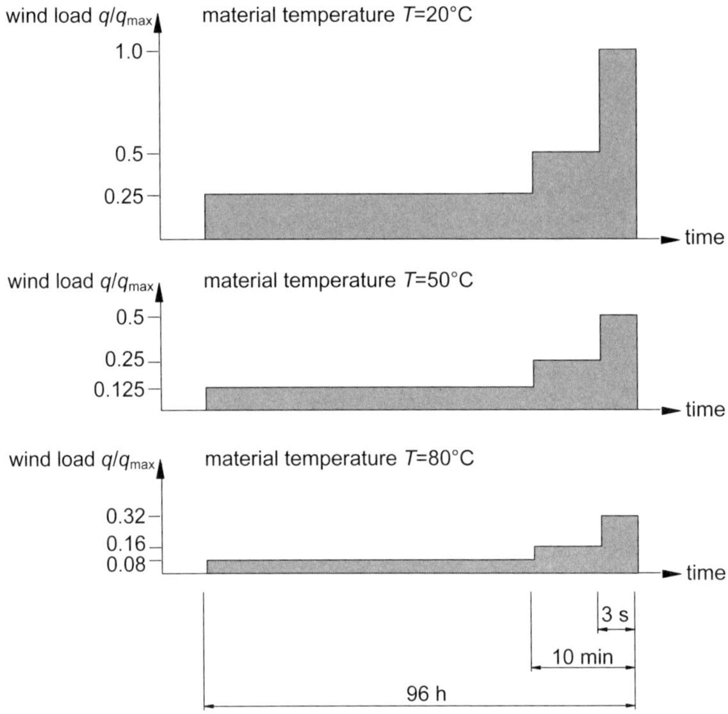

Figure 2.6:
Correlation of wind load and material temperature on façades in Mid-Europe, design load cases [338].

has been determined during one year with measurements on laminated safety glass having a black enamel coating. The contour lines indicate the expected number of days in a period of 100 years that exceed the correlated gust wind load and material temperature.

Wellershoff [338] suggests the use of *Figure 2.6* for Germany and all countries in central continental Europe where the extreme wind situations only occur during storms and cyclones. With a maximum duration of 4 days for cyclones, the figure can be used for the design of viscoelastic materials in façades. In this instance three load cases (20°C, 50°C and 80°C) have to be considered. For each load case the load portions have to be superimposed, taking into account the corresponding load duration.

Further investigations in [338], based on the correlation between maximum wind speed and material temperature, have shown that a simplified single value of

$G_{PVB} = 0.4\,N/mm^2$ may be used to describe the shear modulus of PVB for maximum wind loading.

2.2.4 Seismic loads and movements

Despite recent intensive research activity on the performance of glass in extreme events such as severe windstorms, there is little information in international building codes regarding the seismic design of architectural glass systems. This lack of information is disturbing when one considers the hazards from falling glass and the high costs associated with the loss of building enclosure.

A relatively small but focused research programme carried out in the United States has shown that there are significant differences in performances of various glass types subjected to simulated racking movements. Annealed and heat strengthened laminated glass tends to perform much better in terms of glass fall out when compared to monolithic glass or monolithic glass with unanchored adhesive film. Furthermore silicone bonding of the glass to the frame was found to offer substantial improvements over the alternative dry-glazing details. [34].

2.2.5 Impact loads

Glazing balustrades, glass doors or wall elements should be designed to resist dynamic human impact. As a first approach, human impact loads may be applied on the glass element as a static load, e. g. 1.5 kN at railing height. The loads and the method of application vary between different countries [54, 137]. For non-standard glass elements (e. g. point supported balustrades) and load-bearing partitions (e. g. railings and balustrades) a dynamic analysis or impact tests are often required. The tests specified in this case are either of the impact pendulum type or of the drop ball type and are described in Section 2.4.3 and Section 2.4.4).

2.2.6 Bomb blast

Glass fragments are the primary source of injury in urban explosive events. In the urban environment glass fragments are responsible for 80% of total injuries, and up to 55% of the injuries at 120 m from the blast are glass related [300, 319]. The increasing occurrence and severity of crime and terrorist activities in recent years have therefore significantly increased the need for protective glazing design, i. e. glass façades with enhanced blast performance. The primary purpose of glazing protection is to minimize the number of injuries caused by sharp-edged fragments that are propelled from glazed openings when glass is subjected to blast. The two other aims of glazing protection are to minimize damage to equipment within the building (minimize loss of property) and to allow re-occupation of the building within the shortest period of time (minimize loss of business).

When an explosion occurs, gaseous products of the reaction are formed at very high temperature and pressure at the source. These high pressures expand rapidly into the surrounding medium thus forming a shock wave of compressed air. The shock wave travels radially away from the burst point, gradually reducing its peak pressure, but with increasing duration. The shock wave's leading edge is a near instantaneous rise from

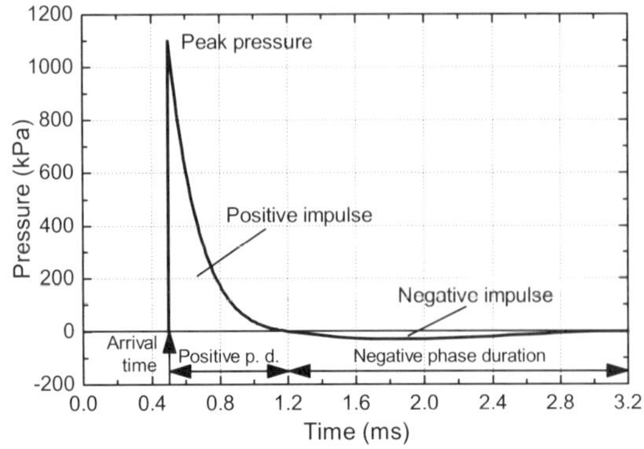

Figure 2.7: Typical free-air blast wave profile [320].

ambient pressure to peak pressure, with a decaying after-edge peak to ambient pressure known as the positive phase. This is followed by a negative pressure period known as the suction phase (*Figure 2.7*). Explosive charges situated extremely close to a building impose highly impulsive, high intensity pressures over a localized region of the structure whereas charges situated further away produce a lower intensity, longer duration pressure distribution over the entire structure. The latter does not normally lead to major structural damage but often causes widespread damage to light cladding and glazing.

The response of a structure to a blast shock wave is greatly influenced by the ratio of the positive phase duration and the natural period of vibration of the structure under consideration. When this ratio is less than 0.2, the effect is considered to be impulsive; where it is higher than 10 it may be considered as a quasi-static load. If the ratio is between 0.2 and 10 the response of the structure is dynamic, and dynamic augmentation of the load may take place. [219].

In addition to the shock wave described above, explosion damage is also caused by the associated movement of air molecules causing a dynamic pressure, often referred to as the 'blast wind'. However, in unconfined explosions the effect of dynamic pressure is much lower than the shock wave pressure and diminishes rapidly with distance from the source. The ground surface and the surrounding buildings may also amplify the blast and therefore have a significant influence on the determination of the design blast load. This third component of the blast load is often referred to as the 'reflected pressure'. Details on blast wave characteristics and types of confined and unconfined explosions are outside the scope of this document. Further information is available in [178, 218, 252]. In practice, the positive exponential pressure-time history of blast waves can be approximated using a triangular impulse load with an equivalent impulse to the exponential pressure-time history, zero or minimal rise time, and linear decay. The impulse parameters may be determined by means of existing computer software [273] or approximate methods shown in [178]. In the past, building owners, employers and designers were obliged to adopt 'reasonable duty of care' by considering the possible effects of blast loading on their buildings. More recently, many countries have adopted legally enforceable regulations to ensure that the building provides adequate safety to occupants when there is a threat of a terrorist act.

The blast loading evaluation should start with a risk analysis of the likelihood of an attack on a building and the identification of consequences due to an explosion [78]. In such an evaluation, the following factors should be taken in account:

- political stability,
- value of the building, its function and the nature of business,
- vulnerability and accessibility of the area,
- the building location and the closeness to possible targets.

In general it is difficult to define a typical explosion scenario as it depends on the type of explosion, the size of the explosive and the location of the attack. Professional advice can be sought from public agencies such as the Police and government departments responsible for national security. A publication by the UK Security Service [301], for instance, provides additional guidelines on threat assessment.

The response of glass to a given blast load depends on:

- the glass panel characteristics (size, thickness, glass type and composition),
- the stiffness and robustness of the frame or support structure,
- the type of connection between the glass panel and the frame.

For the design of façades, two different approaches exist: The first consists of the design for a 'no break' scenario, where glass panels are designed to resist the blast load without breaking. This generally results in very thick and expensive glass configurations set in very stiff and heavy frames. A more common approach is to design the glass for a safe breakage. The latter may be achieved in a number of ways including the combined use of anti-shatter film and bomb blast net curtains or the combined use of fully tempered glass and anti-shatter film. However the most effective approach is to use laminated safety glass (with a PVB interlayer). In this case the blast energy is partially absorbed by the glass breakage and the remaining blast energy is taken by a further deformation of the viscoelastic interlayer. The minimum thickness of laminated glass used for these applications is 7.5 mm including a minimum 1.5 mm thick PVB interlayer. Furthermore the glass should be fixed in a robust frame with a bearing in the order of 30 mm. Bonding with structural silicone tends to improve the overall blast performance. The thickness of glass and the size of the bearing will vary depending on the size of the glass pane. Some initial sizing recommendations are given in [178]. A very effective form of protection can also be provided by a double-glazed unit where a tempered outer glass pane is combined with a laminated inner pane.

Unfortunately, commonly available glass software is unable to simulate the behaviour of broken laminated glass. Consequently, a reliable design without prototype testing is difficult. Blast testing is normally carried out by means of arena testing in a secure range testing site, which involves the glazing prototype mounted onto a test cubicle and located at a standard stand-off distance from a live explosive charge. The explosive charges are measured in TNT equivalent and include 20 kg TNT to simulate hand held bombs and 100 kg TNT to simulate car bombs. The aim of these tests is to assess the hazard consequences by measuring the distance that the glass fragments are projected into the test cubicle (*Figure 2.8*). The hazard levels range from 'A – no break' to 'F – high hazard'.

An alternative form of testing is by employing a shock tube to simulate the blast wave loading on glass panels. This is often used for simulating small and large vehicle bombs.

Current standards related to bomb blast design and testing of windows and façades are EN 13541 [122], EN 13123 [119], and EN 13124 [120, 121], the European code drafts DIS 16934 [217] and DIS 16933 [216] and the US GSA standard [183]. For a review of these and other existing blast testing standards, readers should refer to [223].

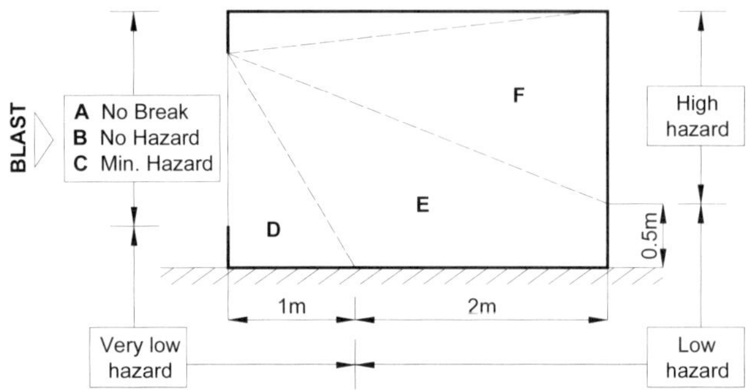

Figure 2.8: Cross-section through blast test cubicle showing hazard ratings.

Some government agencies, particularly those in the UK and USA, have commissioned numerous arena and shock tube tests of PVB laminated glass panes with fully supported edges. These data are generally classified or restricted, however some guideline documents with restricted availability may be provided to bona-fide designers [198]. Most of the existing guidelines and design charts in these guidelines relate to glass panel sizes measuring 1.0 m × 0.8 m and with the four edges fully supported by a clamping frame. There is very little advice or guidelines available for bespoke façade systems or bolted glass. In such cases, full prototype testing is often required.

2.2.7 Internal pressure loads on insulated glass units

Insulated glass units are subjected to internal pressure loads due to pressure difference between the enclosed air cavity and the environment. The pressure difference depends on:

- atmospheric pressure,
- altitude difference between place of fabrication and place of installation,
- variation in temperature,
- the bending stiffness of the glass layers.

Internal pressure loads are particularly critical for small glass dimensions in a façade. The smallest glasses provide the highest lateral plate stiffness and therefore an expansion of the enclosed air results in very high tensile stresses on the plate surface. Similarly high stresses may be generated in curved IGUs where the curvature in the glass results in an increased lateral stiffness that translates into higher surface stresses. Design guidelines for IGUs are provided in prEN 13474-2:2000 [281] and ASTM E 1300-04 [22].

2.2.8 Thermal stress

Thermally induced stresses in glass are generally caused by the presence of a temperature gradient across the glass surface. The source of the heating energy may either be the sun or local heating devices. Glass plates exposed to sunlight are subjected to solar irradiance. A percentage of the incident energy is reflected, some energy is absorbed by the glass and the remaining energy is transmitted through the glass (*Figure 1.16*). The absorbed energy increases the temperature of the glass. In the case of a framed glass or partial shading, only the unshaded areas are exposed to solar energy. The warmer areas expand relative to the cooler ones and generate a tensile stress in the cooler regions of the glass. If the temperature difference between the cooler and the warmer regions is sufficiently high, the stress causes breakage of the glass. In general the risk of thermal breakage is much higher for annealed glass than for HSG or FTG. A common severe condition is on a bright sunny morning after a clear cold night. In this scenario the whole glass is in a cool state when the exposed glass areas are heated rapidly by the solar radiation while the shaded areas (i. e. glass edge behind the glazing bead or frames) remain cool. The heated areas expand and impose a tensile strain on the unheated edges of the plate that acts parallel to the glass edge. The temperature distribution in glass panels is mainly influenced by:

- solar energy absorption of the glass,
- solar radiation intensity,
- heat transfer coefficient,
- heating energy from other energy source i. e. radiators,
- diurnal temperature range,
- internal temperature rise,
- blinds (internal, external),
- shadows.

The strength of glass against thermal stress failure is usually given as an allowable maximum temperature difference. If the calculated temperature difference is less than the allowable temperature difference ΔT_{adm} the panel is thermally safe. There are several existing calculation methods. *Table 2.1* gives an example of maximum allowable temperature differences for different glass types and edge qualities. The values are based on tests carried out by Pilkington in a cooling frame and are derived for an assumed load duration of 3.5 hours per day [70].

The French code [91] additionally provides a calculation method for determining the existing maximum temperature difference by taking into account the parameters listed above. These are compared to the maximum allowable temperature differences that are provided as a function of the glass type, the edge quality and the inclination of the glass panel.

	As-cut or arrised (°C)	**Smooth glass** (°C)	**Polished** (°C)
Float or sheet glass, $h < 12$ mm	35	40	45
Float glass, $h = 15$ mm or 19 mm	30	35	40
Float glass, $h = 25$ mm	26	30	35
Patterned glass	26	26	26
Wired glass	22	22	22
Heat strengthened glass (all types)	100	100	100
Fully tempered glass (all types)	200	200	200
Laminated glass	Smallest value of the component panes		

Table 2.1: Maximum allowable temperature difference ΔT_{adm}.

2.2.9 Surface damage

For non-structural glass elements and for structural elements in which the surfaces are permanently and safely protected, so that they do not undergo any surface damage from external influences, the amount of surface damage acceptable is often controlled by optical acceptance levels such as given in EN 572-2:2004 [149] and ISO 12543-6:1998 [210].

However, many structural glass elements may potentially be exposed to accidental impact, vandalism, heavy wind-borne debris or other factors that result in surface flaws that are substantially deeper than the 'natural' flaws caused by production and handling. Such elements will be called 'exposed glass elements' and the deep flaws 'severe damage' hereafter.

At the instant of damaging the glass surface, the glass is subjected to an elastic stress intensity. If this stress intensity exceeds the fracture toughness, instantaneous failure will occur (see Section 3.3.1). Predicting the crack path or fracture pattern in glass is a complex issue involving dynamic fracture mechanics. A review of the current knowledge in this area is provided in Section 3.4 and [265].

If the instantaneous stress intensity is less than the fracture toughness, some local surface damage may still occur. This damage reduces the strength of the glass element significantly (cf. *Figure 3.6*). When sizing exposed glass elements , the engineer should ideally make a sensible assumption of the potential damage caused by various surface damage hazards (cf. Section 2.1.2). However, there is currently a lack of information on how this qualitative assessment may be translated into quantitative design values. [1] Project specific testing and a considerable amount of engineering judgement are, therefore, generally required. Future research in this field should be conducted in order to establish a relationship between common hazard scenarios and the surface damage that they cause.

2.3 Structural analysis and modelling

In addition to the complex nature of the material strength discussed in Chapter 3, the engineer is also faced with the task of stress and deflection analysis. This adds another layer of complexity to the design, particularly when 'unconventional' support conditions and large deflections are involved. The ensuing sections provide some general guidelines and key references in this regard.

2.3.1 Geometric non-linearity

In contrast to most other building components, glass elements commonly experience large deflections (i. e. in excess of their thickness) prior to failure. In situations where the glass plate is loaded laterally and has translational restraints along its edges, large displacements will cause the mid-plane to stretch thus developing in-plane or membrane stresses that increase the plate stiffness. An increase in plate stiffness may also be observed when the ends of the glass element are not restrained, e. g. when circumferential membrane stresses are set up as the plate is constrained to deform into a non-developable surface.

[1] Some experimental data on the depth of scratches created with a sharp diamond tip are provided in [189].

In these large deflection situations, the assumptions of Kirchhoff's plate theory are violated. Therefore a geometrically non-linear approach that is able to take membrane stresses into account must be used. A mathematical description for the non-linear behaviour is provided by von Karman's partial differential equations (which in the interest of brevity are not reproduced here), however the analytical solution of these equations is complex and unsuitable for manual calculation. Further information on the equations and decoupling solutions are available in specialized literature such as [324]. In practice, it is common to use approximate computational methods, such as the finite difference method or the finite element method, to solve geometrically non-linear problems.

Failure to perform a geometrically non-linear analysis for large deflection situations will result in an overestimation of the lateral deformations. Therefore the actual tensile stresses for a given load are less and the actual tensile stresses for a given deflection are generally greater than those indicated by a linear analysis. An illustration of this non-linear behaviour is provided in *Figure 2.9*. The figure shows the uniform lateral load vs. maximum deformation of a 1676.4 mm × 1676.4 mm × 5.66 mm thick FTG plate, which is linearly supported along all edges. It can be seen that the non-linear finite element analysis provides a reasonably good prediction of this behaviour, however a linear analysis results in gross errors particularly at higher loads.

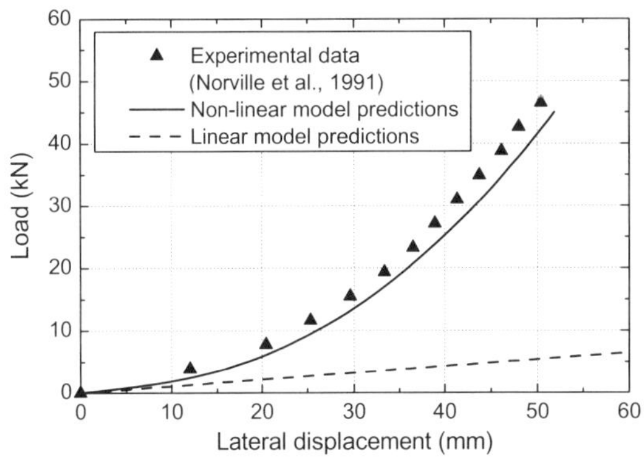

Figure 2.9:
Load vs. displacement relationship for a 1676.4 × 1676.4 × 5.66 mm thick fully tempered glass plate.

2.3.2 Finite element analysis

For common geometries and loading conditions (e. g. a glass panel with simple supports along its edges and subjected to a uniform lateral load), hand calculations based on the tables and graphs in common design standards are usually sufficient for determining maximum stresses and maximum deflections. Unusual glass geometries and support conditions (e. g. curved glass, glass with re-entrant corners, point fixings, non-uniform loading) normally require a more detailed computational analysis.

Various software applications with finite element capabilities are now available and may run on single processor personal computers at relatively little cost. This accessibility and versatility of the finite element method means that virtually every engineering design office has the means to carry out some form of finite element analysis.

Incorrect modelling or misinterpretation of computer-generated results may result in an unsafe estimation of stresses and must therefore be avoided. Detailed advice on the correct use of the finite element method is beyond the scope of this publication and the reader should refer to specialized publications on the subject (e. g. [73]). However, some general rules for the modelling of glass elements are given in the following:

- The mesh density should endeavour to match the expected stress concentrations, i. e. a finer mesh should be adopted around bolt holes and other geometric discontinuities in the glass.
- The results for any given mesh density should be verified by carrying out convergence tests to ensure that any further mesh refinement does not affect the magnitude of the stresses obtained from the analysis.
- Contact between glass and hard materials, such as steel, is normally prevented by using a liner, gasket or bushing that has a lower modulus of elasticity than that of glass (e. g. Nylon, POM, aluminium, EPDM). One important consideration when modelling a fixing region is, therefore, to ensure that the contact surfaces and releases are modelled such that forces are transmitted in compression only and that no tension is transmitted through the gap. This can normally be achieved by using contact elements or by prescribing contact and non-contact surfaces. This approach requires a non-linear analysis.
- Details must be modelled with care. In a point fixing, for instance, the rotational stiffness assigned to the model should match that of the specified bolt, i. e. whether the bolt is free to rotate as in fully articulated bolts or allows only partial rotation as in spring-plate type fixings.

Further advice on the buckling behaviour of glass structures is available in [35, 37, 243] and for the finite element modelling of point fixings see [315].

2.3.3 Simplified approaches and aids

Approximate solutions from tables or graphs provide a quick way to perform maximum stress and deflection calculations for glass panels in flexure [350] and stress concentrations around point supports [275]. These approximate analytical solutions also provide the means for verifying more complex finite element analyses.

Glass selection charts provided in ASTM E 1300-04 [22] and prEN 13474-2:2000 [281] cater for a range of rectangular, circular and triangular flat plates and a variety of support conditions. In the case of a non-rectangular polygonal shape, an approximation of the stresses may be obtained by exercising some engineering judgement as suggested by Vallabhan et al. [333]. This method involves representing the polygonal glass element with an equivalent circular pane that circumscribes the polygonal plate. The maximum stresses and deflections for the equivalent circular plate may be obtained from [350].

2.4 Requirements for application

This text has been compiled in collaboration with the following experts:
Dr. Iris MANIATIS, Prof. Dr. Geralt SIEBERT

Besides the fundamental issues of load carrying capacity and deflections limits, there are often additional performance requirements for structural glass elements in façades, roofs or floors. Many of these additional requirements are not fully covered by standards and are often too complex to verify numerically. Therefore these performance requirements, such as the post-breakage structural capacity and the impact resistance, often require full-scale testing. This section provides an overview of the more important considerations and current regulations on this subject.

Most building codes require that a building and its elements must be designed and built so as to reduce the risk to public health and safety to societally acceptable levels. This requirement can in general be guaranteed in three different ways:

- For *standard* applications by applying technical rules and design codes.
- For *frequent* applications by applying *European technical approvals* (ETA) or *national general approvals.*
- For *individual* applications by obtaining *special permits* given by certain authorities and in some countries by *engineering judgement* of the responsible designer.

Where structural elements are regarded as being safe or harmless, no special requirements apply.

2.4.1 Vertical glazing

Any glazing with an inclination of less than 10 or 15° to the vertical (depending on the country) is considered as 'vertical glazing'. If vertical glazing acts as an anti-drop-device, it has to fulfill additional requirements that are discussed in Section 2.4.4. The present paragraph concentrates on building façades and sound-screens.

For façades, ensuring structural safety is often sufficient. For structural sealant glazing systems (SSGS), the requirements in the European Technical Approval Guideline (ETAG) No. 002 [163] must be met. In some countries, additional national regulations may apply (e. g. additional tests for hurricane resistance are required by some states in the USA). There are usually no specific requirements for the post-breakage resistance of vertical glazing and consequently there is no restriction on the type of glass that may be used in practice:

- **Europe — prEN 13474 [280, 281].** New European standards are currently developed and are meant to take the current state of knowledge into account. Design is based on the partial factor approach as it is used in other European codes. In its final version, it should provide the basis for almost all glass applications and provide detailed sizing guidelines for glass with simply supported edges.
- **Germany.**
  - *DIN 18008 [81, 82].* National standard under development. Aims and scope are very similar to the above-mentioned European draft standard prEN 13474.
  - *TRLV 2006 [328].* This document contains regulations for linearly supported overhead and vertical glazing. In addition to design rules, allowable tensile stress and deflection limits, a method for calculation of climatic loads in insulating glass units is also provided.
  - *ZTV-Lsw 88 [353].* These are regulations for sound screens that were published by the German Federal Ministry for Traffic and cater for sound screens along roads constructed of any material. The main requirement concerning the application of glass, is a minimum thickness of 12 mm for fully tempered glass and 16 mm for laminated glass elements.

- **USA — ASTM E 1300**. The American National Standard ASTM E 1300 [22] applies to vertical and sloped (overhead) glazing in buildings exposed to a uniform lateral load. The types of glass covered by this standard include monolithic, laminated, or insulating glass of rectangular shape with simple supports along two, three or four edges. The specified design loads may consist of wind load, snow load and self-weight with a total combined magnitude $\leqslant$ 10 kPa. It does not apply to other applications such as balustrades, glass floor panels and point-supported structural glass members. Useful deflection limits are also provided.
- **United Kingdom — BS 6262**. Part 3 of this British standard [51] provides glass thickness selection charts for a range of lateral wind loads and caters to annealed, fully tempered, insulated glass units. The support conditions are limited to rectangular glass panels that are simply supported along the four edges and panels. Part 6 of this standard [53] gives good practice detailing and basic sizing recommendations on special applications such as glass fins and simple bolted glass barriers and glass partitions.

2.4.2 Overhead glazing

Glazing with an inclination of more than 10 or 15° to the vertical is considered as overhead glazing, because people might step under it. If the glazing is used as a walking surface for either regular access or occasional maintenance, it is often referred to as 'accessible glazing' and is discussed in (cf. Section 2.4.3). Overhead glazing elements must stay in position for a certain amount of time after breakage to minimize the likelihood of injury (cf. Section 2.1.3). Design of overhead glazing involves structural safety *and* sufficient post-breakage resistance. The latter can be proven by testing (cf. Section 8.1) or ensured by additional features that prevent broken glass from falling down (e. g. steel ropes or nets under the glass elements). If the glazing is accessible for maintenance, more onerous requirements come into effect, namely that the broken glass is able to support the weight of a person for a specific amount of time after failure.

Monolithic fully tempered glass elements should not be used for overhead applications as they do not provide any residual integrity or resistance after breakage. For laminated safety glass made of fully tempered glass sheets, the residual resistance has to be proved by testing (cf. Section 8.1). Laminated glass made of annealed or heat strengthened glass is generally able to satisfy normal post-breakage requirements as long as the glass panel is within a given set of spans and support conditions (e. g. the span of a glass panel with two linear supports along its opposite edges should not exceed 1200 mm). In case of point supported glass elements, the residual resistance is strongly influenced by the type of point fixing. Countersunk point fixings usually show a poor residual resistance whereas point fixings that exert some clamping force on the glass and the interlayer (e. g. laminated glass with two-sided raised heads) perform better. The following regulations are often used in practice:

- **Germany — TRLV**. TRLV 2006 [328] applies only to linearly supported glazing. It does not cover glued façade panels, glazing acting as bracing elements or bent overhead glazing. Railings or accessible glazing have to comply with additional requirements. Greenhouses and dormer windows up to a size of 1.6 m^2 in private dwellings do not need to conform to TRLV. The glass types that may be used

are annealed glass, cast glass, heat strengthened glass, fully tempered glass and laminated safety glass built up from the above mentioned glass types and a PVB interlayer, or an alternative foil or cold poured resin.
Deflection of bearing elements must be limited to 1/200 of the span and 15 mm. They have to support the glass panes for loads *and* wind suction. For single glazing elements and the lower pane of insulating glass elements, only wired glass or laminated safety glass made of annealed or heat strengthened glass is allowed. FTG must not be used for such applications.

Glass elements with a span exceeding 1.20 m have to be supported on all edges. It is important to note that if the aspect ratio exceeds 3:1, the glass pane is considered to be supported along the two longer edges only. The minimum thickness of PVB is generally 0.76 mm; 0.38 mm is allowed for glass elements with spans not exceeding 0.8 m and that are supported along all four edges. No notches or holes are allowed in overhead glazing. Positive effects due to shear transmission by the PVB interlayers of laminated glass elements or due to edge sealing of insulating glass units may not be taken into account.

Structural verification is based on maximum allowable stresses and deflections, see *Table 4.1*.

- **USA — ASTM E 1300**. See Section 2.4.1.
- **United Kingdom**. There are no mandatory requirements in the UK for overhead glazing as long as the glass is not used as a walking surface, including for maintenance and cleaning [77]. Common practice in this case is to provide some form of safety glass, preferably laminated glass [76]. In cases where objects may fall onto the glass surface, hard body impact tests should be performed as specified in EN 356:1999 [146].

2.4.3 Accessible glazing

Floors, roofs and other horizontal glazing are often either accessible to the public or at least accessible for cleaning and maintenance. Resistance against impact caused by a hard or soft body as well as the post-breakage behaviour must therefore be examined as well as slip resistance of the glass surface. In order to satisfy these conditions full scale tests are often required. Currently, there is no European standard for the design of accessible glazing. The following sections therefore discuss some available and useful recommendations and rules.

- **Germany**. A recommendation by the German Institute for Building Technology [79] is commonly used.
  For *glazing with public access*, it contains the following requirements:
  - **Structural design**. For the walking surface, laminated safety glass made of at least three glass panes should be used and tied together by PVB interlayers. Linear support as well as point fixing with or without drilled holes is acceptable. For basic design requirements, TRLV 2006 [328] is applicable. Structural analysis must be performed by considering all influences that may increase the stresses, like temperature deformations, supports, eccentricities, tolerances etc. The uppermost glass sheet of laminated glass elements must be neglected for the structural analysis. Furthermore, composite behaviour (shear transmission

through the PVB) may not be taken into account. The deflection limit at midspan is 1/200 of the span.

◇ **Impact resistance**. Impact resistance has to be verified by tests, see Section 8.1.3.

◇ **Residual resistance after breakage (remaining load carrying capacity).** For this test, the damaged specimen from the impact test is used. Any glass sheets within the laminated glass element that were not destroyed by the impact have to be damaged with a hammer and center punch at several points. The post-breakage resistance is defined as the time elapsed between the breaking of all the glass sheets and the collapse of the loaded laminated glass element. Normally, a post-breakage resistance of at least 24 hours is required by German authorities (cf. Section 8.1).

For glazing that is accessible for *cleaning and maintenance,* basically the same requirements apply. The applied load for the impact test however is lower and the impact body is different. Details are given in Section 8.1.3. Additional safety requirements for permanently installed working and walking areas and other installations for maintenance and inspection are given in DIN 4426:2001 [84].

- **USA — IBC**. Glass requirements are found in Chapter 24 of the International Building Code (IBC) [200]. Section 2409 covers glass in floors and sidewalks. Laminated glass with a minimum of two sheets has to be used for such applications. There are no requirements concerning impact resistance or post-breakage resistance.
- **United Kingdom**. If access onto the glass is required , the Health Safety and Welfare regulations come into effect and testing for both soft and hard body impact is mandatory. The post-breakage performance of the glass is tested by the 'sand bag' test. Advice on the test regimes and selection of the glass is available in [77] and [175].

2.4.4 Railings and balustrades

Impact resistance, especially against human impact, is a key requirement for railings and balustrades. Several accidental hazard scenarios have to be considered:

- persons falling down or slipping through gaps in the barrier;
- injury due to glass fragments or sharp edges after glass breakage;
- injury by glass fragments falling on areas below that may be occupied by people.

Railings and balustrades can be divided into three groups:

- Category A — Full height glazing. Mostly glazing with continuous lateral support and without handrail that acts as a wall or forms part of a wall.
- Category B — Cantilevered balustrade. The glass panels are clamped along their lower edge. The handrail is attached to the upper edge.
- Category C — Balustrade with a glass infill panel. The glass infill panels may be fixed by continuous lateral support along at least two edges or by point fixings. In some countries, clamped or clipped infill panels are also possible.

A selection of international and national regulations on balustrades and railings are:

- **Europe — EN 12600**. Impact resistance requirements for flat glass in buildings (e. g. railings and balustrades) are given in the European standard EN 12600:2002 [105]. The test simulates human body impact using a 50 kg mass wrapped by two rubber tires (soft pendulum test). The test is intended to classify flat glass products according to their impact resistance and mode of breakage (cf. Section 8.1). However, the standard does not specify requirements for application therefore, various additional national requirements exist in European countries. In most countries (in contrast to Germany), impact resistance has to be verified for the glazing rather than the whole assembly. This means that the glass may be tested in a standardized frame instead of using the original components that will be used in the building.
- **Germany — TRAV**. According to TRAV 2003 [327], the impact resistance of the glass and its supporting system (including clamps, point fixings) has to be verified in addition to standard structural design. Different glass types have to be used in function of the balustrade category (A to C). There are two different methods for impact resistance verification:
  - *Soft pendulum test*: As distinct from EN 12600:2002 [105], the experimental setup and the specimen have to be equivalent to the original building unit in terms of materials, support structure etc. The same standard pendulum is used.
  - *Verification by calculation*: TRAV 2003 [327] contains tables with maximum allowable stresses for various glass dimensions and thicknesses. These may be used for railings and balustrades of categories A and C (continuous lateral support along at least two edges). Furthermore, TRAV gives detailed specifications of many standard railing and balustrade types with verified impact resistance. The use of these standard details ensures conformity without further verification.
- **UK — BS 6180**. BS 6180:1999 [49] contains requirements for barriers in and around buildings. Actions have to be determined according to BS 6399-1:1996 [54]. Structural safety of all balustrade components must be verified. Different glass types have to be used in function of the balustrade type. For impact performance, the safety glazing recommendations in BS 6262-4:1994 [52] have to be taken into account. Barriers with glass infill or cantilevered balustrades must comply with impact classes. This is verified by soft pendulum tests according to BS 6206:1981 [50] (see Section 8.1.3).
- **USA — IBC**. Glass requirements are covered in Section 2407 'Glass in handrails and guards' of the International Building Code (IBC) [200]. Actions have to be determined according to Section 1607.7 of the IBC. Different glass types have to be used in function of the balustrade type. The impact class is determined by soft pendulum tests according to CPSC 16 CFR 1201 [74] (see Section 8.1.3).

Chapter

3 Fracture Strength of Glass Elements

3.1 Introduction

The aim of this chapter is to provide an in-depth understanding of the mechanisms of glass fracture that underpin subsequent chapters and should be used as the basis for structural design of glass.

The mechanical properties of glass stem from the molecular structure discussed in Chapter 1, which unlike most other construction materials, does not consist of a geometrically regular network of crystals, but of an irregular network of silicon and oxygen atoms with alkaline parts in between. The random molecular structure has no slip planes or dislocations to allow macroscopic plastic flow before fracture; consequently, glass is perfectly elastic at normal temperature and exhibits brittle fracture. This inability to yield plastically before fracture means that the fracture strength of glass is very sensitive to stress concentrations. Since surface flaws cause high stress concentrations, accurate characterization of the fracture strength of glass must incorporate the nature and behaviour of such flaws. To this end, Section 3.2 discusses the stress corrosion that causes existing surface flaws to grow slowly in size prior to failure, a phenomenon that is often referred to as 'sub-critical crack growth'. This section is also a prerequisite for subsequent sections.

Section 3.3 introduces quasi-static linear elastic fracture mechanics (LEFM) and provides a mathematical model for determining the fracture strength of glass. This model, called the 'lifetime prediction model', is derived from a mathematical description of a glass element's surface condition and of the growth and fracture of surface flaws through LEFM and probability theory. The equations that are provided in the lifetime prediction model can be used for predictive modelling and structural design. They take sub-critical crack growth, non-homogeneous, time-variant biaxial stress fields, arbitrary geometry and arbitrary stress histories into account. While the lifetime prediction model described herein is more complex than traditional semi-empirical models, it offers significant advantages that are discussed in this section.

Although it is valid for very short loading times, the lifetime prediction model in Section 3.3 cannot be used to describe dynamic phenomena such as glass fracture or the response of glass elements to impact loads. To deal with such phenomena, dynamic fracture mechanics theory is required. Most aspects of this theory are of formidable theoretical complexity and beyond the scope of this document. However, some of the simplified empirical formulations are of practical interest especially for the diagnostic interpretation of glass failures and are therefore presented in Section 3.4.

Because most of the glass used in construction is soda lime silica glass, the present chapter refers to this glass type. The presented concepts and mathematical relations are also applicable for other glass types, but the material parameters need to be adjusted (cf. Chapter 1).

3.2 Stress corrosion and sub-critical crack growth

In vacuum, the strength of glass is time-independent.[1] In the presence of humidity, however, stress corrosion causes flaws to grow slowly when they are exposed to a positive crack opening stress. This means that a glass element which is stressed below its momentary strength will still fail after the time necessary for the most critical flaw to grow to its critical size at that particular stress level. The momentary strength of a loaded glass element therefore decreases with time, even if it is exposed to static loads only. This phenomenon, which is fundamental for the structural use of glass, was already discovered in 1899 by Grenet [180]. The growth of a surface flaw depends on the properties of the flaw and the glass, the stress history that the flaw is exposed to and the relationship between crack velocity and stress intensity.

In the present document, the term '*stress corrosion*' is used to refer to the chemical phenomenon. The term '*sub-critical crack growth*' is used to refer to the *consequence* of stress corrosion, i. e. the growth of surface flaws.[2]

3.2.1 Relationship between crack velocity and stress intensity

First, systematic investigation of stress corrosion was conducted by Levengood [239]. An explanation for the chemical process behind the phenomenon was put forward by Charles and Hillig [65] and further developed by Michalske and Freiman [255]. This theory, also known as the 'classical stress corrosion theory', involves the chemical reaction of a water molecule with silica at the crack tip (*Figure 3.1*):[3]

$$\text{Si-O-Si+H}_2\text{O} \quad \rightarrow \quad \text{Si-OH+HO-Si} \tag{3.1}$$

According to this theory, the crack velocity scales with the kinetics of this chemical reaction. Its activation energy depends on the local stress and on the radius of curvature at the

[1] Even in vacuum, the resistance of many glasses is in fact slightly time-dependent. This effect, called 'inert fatigue', is however of no practical relevance for structural engineering applications.

[2] In academic publications, this distinction is not always made and other terms, such as 'slow crack growth', 'static fatigue', and 'environmental fatigue' are in use.

[3] The classical interpretation involving a chemical reaction at the very tip of a crack is questioned by Tomozawa [326]. As the diffusion of molecular water into the glass is activated by stress, he suggests that this diffusion process and the modification of the glass properties in the crack tip area that it causes might explain sub-critical crack growth. A more in-depth discussion is beyond the scope of this document. The interested reader should refer to specialized texts on this subject, e. g. [57, 172, 186, 255, 326, 343].

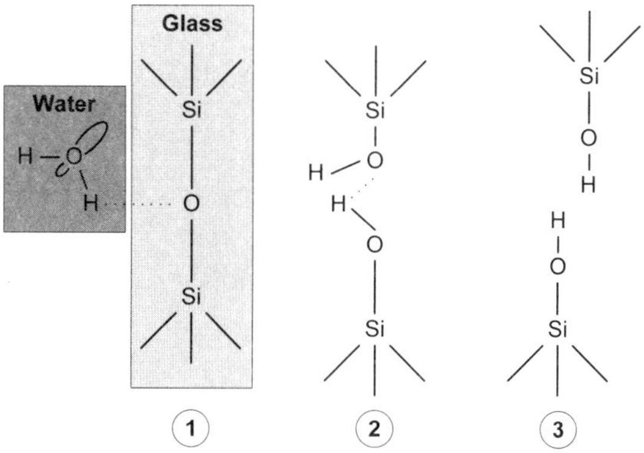

Figure 3.1:
Stress corrosion, chemical reaction at the crack tip: (1) adsorption of water to Si–O bond, (2) concerted reaction involving simultaneous proton and electron transfer, and (3) formation of surface hydroxyl groups [255].

crack tip. The theory involves a first order chemical process, which is consistent with the observed linear correlation between the logarithm of the crack velocity v and the logarithm of the humidity ratio H (except for very low H or v) [343].

Figure 3.2 shows the simplified, schematic relationship between *crack velocity* v and *stress intensity factor* K_I that is commonly used for glass lifetime prediction. For values of K_I close to the fracture toughness K_{Ic} (definition → p. 57) or even above, v is independent of the environment and approaches a characteristic crack propagation speed (about 1 500 m/s for soda lime silica glasses) very rapidly. In a narrow region below K_{Ic} (region III), the curve is very steep, v lying between 0.001 m/s and 1 m/s. In inert environments (cf. Section 3.3.3), this curve would extrapolate linearly to lower crack velocity. In normal environments, the behaviour strongly depends on the environmental conditions. The empirical relationship

$$v = S \cdot K_I^n \tag{3.2}$$

which was originally proposed by Evans and Wiederhorn [165], provides a good approximation for region I.[4] The parameters S and n need to be determined from experiments. The unit of S depends on the value of n. This can be avoided with the following equivalent formulation ($S = v_0 \cdot K_{Ic}^{-n}$):

$$v = v_0 \left(K_I / K_{Ic}\right)^n \tag{3.3}$$

The crack velocity parameter v_0 has the units of speed (length/time), n is dimensionless. When the v-K_I-curve is plotted on logarithmic scales, v_0 represents its position and n its slope. K_{Ic} is a material constant that is known with a high level of precision and confidence (cf. Section 3.3.1). Regions I and III are connected by region II. In this region, the kinetics of the chemical reaction at the crack tip are no longer controlled by the activation of the chemical process, but by the supply rate of water. It takes time for a water molecule to be transported to the crack tip, such that a shortage in the supply of water occurs as the crack velocity increases [343]. The crack velocity v is, therefore, essentially independent of K_I but depends on the amount of humidity in the environment. Below a certain *threshold stress intensity* K_{th} (see Section 3.2.2), no crack growth occurs.

[4] The exponential functions $v = v_i \cdot e^{\beta K_I}$ and $v = v_i \cdot e^{\beta(K_I - K_{Ic})}$ were also proposed to model the v-K_I relationship. In practice, the difference between a power law with a high exponent and an exponential function is very small. An exponential function has the main advantage of being consistent with the kinematics of the above-mentioned chemical reaction. Equation (3.2), however, allows for much simpler calculations, which explains its predominant use.

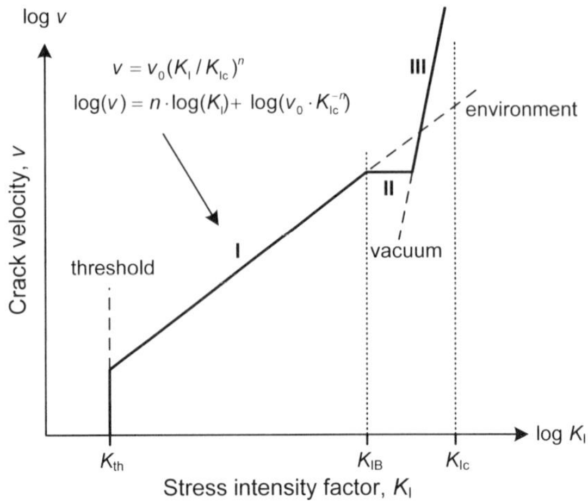

Figure 3.2:
Idealized v-K_I-relationship.

In view of the order of magnitude of glass elements in buildings (mm to m), the typical depth of surface flaws (μm to mm) and the service life generally required, only the range of extremely slow sub-critical crack growth, region I, is relevant for determining the design life of a glass element. The contribution of regions II and III to an element's lifetime is negligible.

3.2.2 Crack healing, crack growth threshold and hysteresis effect

In 1958, Levengood [239] found that ageing has an effect on glass surface flaws. Further experimental work in laboratory conditions showed that the strength of flawed specimens increases during stress-free phases [323, 341]. Looking at it in more detail, this effect, generally called *crack healing*, is a consequence of two phenomena, the crack growth threshold and the hysteresis effect.

At stress intensities below the *crack growth threshold*[5] K_{th}, no significant crack growth occurs. For typical soda lime silica glass at a moderate pH value, K_{th} is about 0.2 to 0.3 MPa $m^{0.5}$ (see Haldimann [189] for an overview of available data).

The crack growth threshold was originally explained by a rounding of the crack tip ('*crack tip blunting*') at slow crack velocities [28, 65]. More recent investigations, however, strongly support the hypothesis that alkali are leached out of the glass and that this change in the chemical composition at the tip of the crack is responsible for the crack growth threshold rather than a geometrical change (blunting). Observations of aged indentation cracks by atomic force microscopy did not give any evidence of blunting. Sodium containing crystallites were actually found on the surface of glass close to the tip of the indentation crack. This is more consistent with alkali ions' migration under the high stress at the crack tip and their exchange with protons or hydronium ions[6] from the environment [173, 185, 259].

In alkali containing glasses, there is also a *hysteresis effect*: an aged crack will not repropagate immediately on reloading. The hysteresis effect is convincingly explained by renucleation of the aged crack in a plane different from the original one, as if the path of

[5]Also known as 'stress corrosion limit', 'crack growth limit', 'threshold stress intensity', or 'fatigue limit'.
[6]A hydronium ion is the cation H_3O^+ derived from protonation of water.

the crack has to turn around the area just in front of the former crack tip. This non-coplanar re-propagation was directly observed by atomic force microscopy [197, 340, 344].

A comprehensive probabilistic crack propagation model, which accounts for the above-mentioned effects, was proposed by Charles et al. [66].

Although its favourable influence can be considerable, to this day, crack healing has not been taken into account (at least not explicitly) by design proposals. Because of the strong dependence on the environmental conditions, crack healing is difficult to quantify. The threshold appears to depend strongly on the environmental conditions and on the glass's chemical composition. It is, for instance, more easily evidenced with alkali containing glasses and in neutral or acidic environments, while there is no evidence of a threshold in alkaline environments [179]. In static long-term outdoor tests, in contrast to tests in the climatic chamber, no evidence of any substantial crack healing or of a crack growth threshold was found [169]. *For structural applications, in which safety is a major concern, it therefore remains advisable not to take any threshold or healing effects into account.*

3.2.3 Influences on the relationship between stress intensity and crack growth

It is important to bear in mind that the relationship between stress intensity and crack velocity is very sensitive to a number of aspects. A short overview is given in the following text. For more details, see [189].

- **Humidity**. As mentioned before, the water content of the surrounding medium[7] strongly influences sub-critical crack growth. The effect of an increasing water content is essentially a parallel shift of regions I and II of the v-K_I relationship towards higher crack velocities [342].
- **Temperature**. An increasing temperature causes mainly a parallel shift of the curve towards higher crack velocities. Furthermore, the slope decreases slightly [343].
- **Corrosive media and pH value**. The crack velocity generally increases as the pH value of the surrounding medium increases. Furthermore, the pH value has a certain effect on the slope of the v-K_I relationship and a particularly strong influence on the crack growth threshold K_{th} [172].
- **Chemical composition of the glass**. All parameters of sub-critical crack growth are influenced by the chemical composition of the glass [343].
- **Loading rate.** According to Haldimann [189], the v-K_I relationship does not only depend on environmental conditions, but is also strongly loading rate dependent. As mentioned before, stress corrosion requires humidity. If an element is loaded rapidly, the diffusion process is not fast enough, so that a shortage in the supply of water to the crack tip slows down stress corrosion and therefore the sub-critical growth of flaws. Consequently, the v-K_I relationship of an element is shifted towards lower crack velocities when loaded rapidly.

Figure 3.3 gives an overview of published v-K_I-data[8] [44, 94, 186, 197, 289, 290, 303, 307, 329, 343]. The following can be concluded:

[7]It is actually the ratio of the actual partial pressure to the partial pressure at saturation. In air, this corresponds to the relative humidity.

[8]When modelling sub-critical crack growth, the v-K_I-relationship is generally assumed to be valid over the full K_I-range. This is why the curves that represent design models extend to the entire range of the figures' axes.

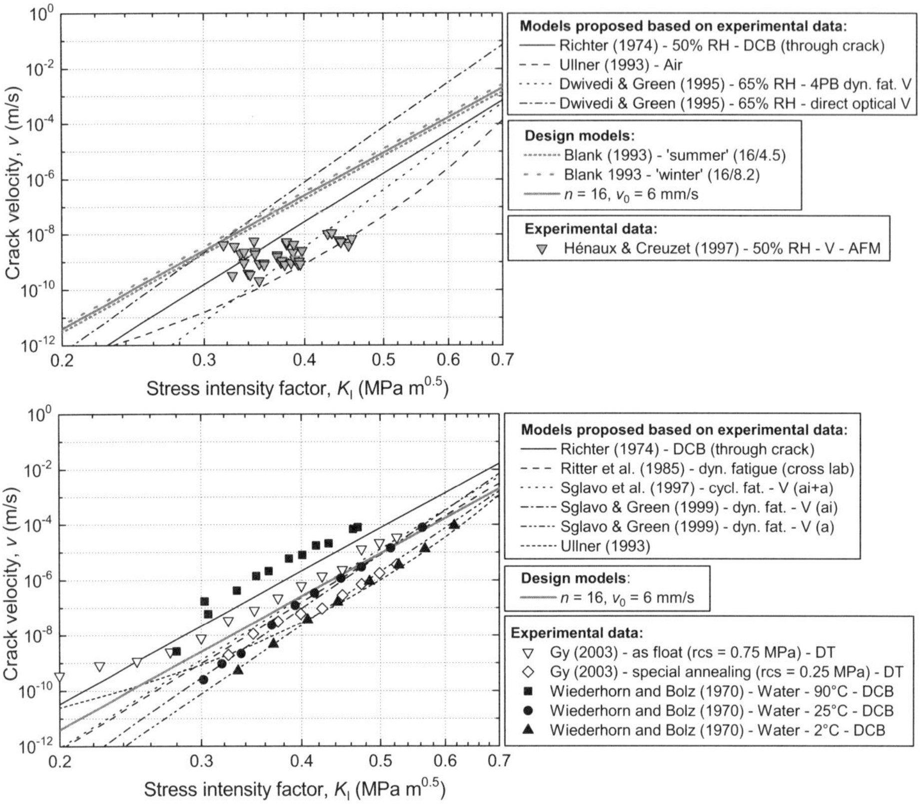

Figure 3.3: Crack growth data overview in air (above) and in water (below).
(V = Vickers indentation, ai = as indented, a = annealed, DT = double torsion test, DCB = double cantilever beam test, dyn. fat. = dynamic fatigue, rcs = residual core stress.

- **General.** Crack velocity parameters vary widely and depend on several influences, including environmental conditions and the loading rate. Fracture strength predictions for service lives of many years are, therefore, of limited accuracy.
- **Structural design.** *For structural design*, a constant value of $n = 16$ is a reasonable assumption. For general applications, $v_0 = 6\,\mathrm{mm/s}$ should be conservative[9]. For glass elements that are permanently immersed in water, a higher value of e. g. $v_0 = 30\,\mathrm{mm/s}$ is more appropriate.
- **Interpretation of experiments.** Strength data from tests at ambient conditions are inevitably dependent on the surface condition *and* on crack growth behaviour. The large variability of the crack velocity parameters makes it very difficult to obtain accurate surface condition information from tests at ambient conditions [189]. Inaccurate estimation of the crack velocity during testing can yield unsafe design parameters. Testing at *inert* conditions is, therefore, preferable (Section 6.4).

[9] Further differentiation of environmental conditions, e. g. considering summer and winter conditions, is not recommended for modelling purposes. The potential difference between the two cases is very small compared to the scatter of the data. The definition of two parameter sets would therefore be rather arbitrary and would not necessarily increase the accuracy of the model. The complexity of the calculation process, on the other hand, would be increased considerably.

3.3 Quasi-static fracture mechanics

Linear elastic fracture mechanics (LEFM) provides a good model for describing the brittle fracture of glass (see Section 3.1). In LEFM, mechanical material behaviour is modelled by looking at *cracks*. A crack is an idealized model of a flaw having a defined geometry and lying in a plane. It may either be located on the surface (*surface crack*) or embedded within the material (*volume crack*). For structural glass elements, only surface cracks need to be considered. *Figure 3.4* shows the fundamental terms used to describe such cracks.

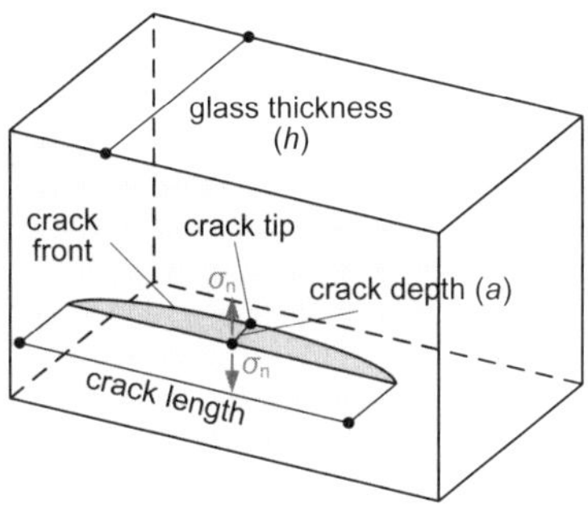

Figure 3.4:
Fundamental terms used to describe surface cracks.

3.3.1 Stress intensity and fracture toughness

The theoretical strength of a material is determined by the forces of the interatomic bonds. *Orowan* proposed that the stress necessary to break a bond, known as Orowan stress, is given by

$$\sigma_{\mathrm{m}} = \sqrt{E\gamma/r_0} \tag{3.4}$$

where γ is the fracture surface energy, r_0 is the equilibrium spacing of the atoms and E is Young's modulus. With $E = 70\,\mathrm{GPa}$, $r_0 = 0.2\,\mathrm{nm}$ and $\gamma = 3\,\mathrm{J\,m^{-2}}$, we obtain a theoretical strength of 32 GPa for a typical silica glass [310]. In practise the tensile strength of annealed soda lime silica glass is much lower. The large variations between theoretical and practical strength were explained by Griffith [181], whose experiments on glass form the basis of modern fracture mechanics. Griffith argued that fracture did not start from a pristine surface, but from pre-existing flaws, known as 'Griffith flaws', on that surface. Such flaws are not necessarily visible to the naked eye, but they severely weaken brittle solids because they produce very high stress concentrations. As explained in Section 3.2, surface flaws in glass grow with time when loaded, the crack growth rate being a function of several parameters.

In 1913, Inglis [201] recognized that a slot, notch or hole in a metal plate tends to reduce its tensile strength by an amount that is more than would result simply from the reduction in load-bearing cross-sectional area. He demonstrated that the stress magnification near the tip of a narrow elliptical discontinuity whereof the long diameter $2a$ lies perpendicular to the applied stress σ_E may be approximated by

$$\sigma_{\mathrm{tip}} = 2\sigma_E\sqrt{a/\rho}\,, \tag{3.5}$$

where ρ is the radius of curvature at the crack tip. Clearly, with atomically sharp flaws, ρ is very small and σ_E is thus magnified by several orders of magnitude such that σ_{tip} can approach the molecular bond strength even if the applied stresses are relatively small.

Based on Inglis' work and experiments on glass specimens, Griffith [181] modelled a static crack as a reversible thermodynamic system. In the configuration that minimizes the total free energy of the system, the crack is in a state of equilibrium and thus on the verge of extension. The total energy U in the system is

$$U = U_{\text{M}} + U_{\text{S}} \tag{3.6}$$

where U_{M} is the mechanical energy (the sum of the strain potential energy stored in the elastic medium and the potential energy of the outer applied loading system) and U_{S} is the free energy expended in creating new crack surfaces. Therefore, U_{M} favours crack extension, whereas U_{S} opposes it. The equilibrium requirement $\mathrm{d}U/\mathrm{d}c = 0$ is known as the *Griffith energy-balance concept*. From this, Griffith calculated the critical conditions at which instantaneous failure occurs as

$$\sigma_{\text{f}} = \sqrt{2E\gamma/(\pi a_c)} \tag{3.7}$$

where σ_{f} is the failure stress and a_{c} is the critical crack length.

Irwin [203] extended the original Griffith energy-balance concept to provide a means of characterizing a material in terms of its brittleness or fracture toughness. He introduced the concept of the stress intensity factor (SIF) K, which represents the elastic stress intensity near the crack tip. The stress intensity factor for mode I loading[10], K_{I}, is given by[11]

$$K_{\text{I}} = Y \cdot \sigma_{\text{n}} \cdot \sqrt{\pi a} \tag{3.8}$$

where σ_{n} is the nominal tensile stress normal to the crack's plane, Y is a correction factor, and a represents the size of the crack (i. e. the crack depth or half of the crack length).[12]

The correction factor Y[13] depends on the crack's depth and geometry, the specimen geometry, the stress field and the proximity of the crack to the specimen boundaries. While the dependence on the specimen geometry, the stress field and the crack depth is small for shallow surface cracks and can generally be ignored, the dependence on the crack geometry and the proximity to boundaries is more significant. Y is therefore often called the *geometry factor* . A long, straight-fronted plane edge crack in a semi-infinite specimen has a geometry factor of $Y = 1.12$. For half-penny shaped cracks in a semi-infinite specimen, the geometry factor is in the range of 0.637 to 0.713, depending on the approach used [189].

Instantaneous failure of a glass element occurs when the elastic stress intensity K_{I} due to tensile stress at the tip of one crack reaches or exceeds a critical value. This critical

---

[10] Opening mode, i. e. normal separation of the crack walls under the action of tensile stresses.

[11] In scientific publications, the energy release rate G is often used instead of the stress intensity factor K_{I}. For an elastic material and crack mode I, it is $G = K_{\text{I}}^2/E'$ with $E' = E$ for plane stress state and $E' = E/(1-\nu^2)$ for plane strain state [203]. E is Young's modulus and ν is Poisson's ratio. In the case of shallow surface cracks, a plane stress state may be assumed.

[12] Such simple stress intensity considerations are based on the assumption of a homogeneous distribution of the nominal stress within the section. Although this is generally not the case in structural glass applications, the equations are good approximations because the depth of the cracks is small compared to the material thickness, such that the stress variation over a crack's depth is small.

[13] Caution when using published data, where Y is often used as a synonym for $Y\sqrt{\pi}$.

value is a material constant known as the plane strain *fracture toughness* or the *critical stress intensity factor* K_{Ic}. This failure condition is called *Irwin's fracture criterion* and is expressed as:

$$K_{\mathrm{I}} \geq K_{\mathrm{Ic}} \tag{3.9}$$

The criterion assumes pure mode I fracture of a crack exposed to uniaxial tension normal to the crack's plane. More general cases are discussed in Section 3.3.5.

The fracture toughness K_{Ic} can be considered to be a material constant. It does not depend significantly on influences other than the material itself. *Table 3.1* gives an overview of published values for modern soda lime silica glasses. A value of $K_{\mathrm{Ic}} = 0.75\,\mathrm{MPa\,m^{0.5}}$ can be used for all practical purposes.

Source	K_{Ic} **(MPa m$^{0.5}$)**
Wiederhorn [342]	0.82
Atkins and Mai [24]	0.78
Gehrke et al. [172]	0.78
Menčík [254]; from a review of published data	0.72–0.82
Ullner [329]	0.76

Table 3.1: Fracture toughness K_{Ic} of soda lime silica glass at room temperature.

3.3.2 Heat-treated glass

The term *heat-treated glass* includes *any* glass type that has been processed in order to induce residual stresses (cf. Section 1.3.2), namely heat strengthened glass and fully tempered glass.

The in-plane surface stress normal to a crack's plane (index 'n'), also known as the *crack opening stress*, is:

$$\sigma_{\mathrm{n}}(\tau, \vec{r}, \varphi) = \sigma_{E,\mathrm{n}}(\tau, \vec{r}, \varphi) + \sigma_{\mathrm{r,n}}(\vec{r}, \varphi) + \sigma_{\mathrm{p,n}}(\tau, \vec{r}, \varphi) \tag{3.10}$$

σ_E surface stress due to actions
σ_{r} residual surface stress due to tempering ('prestress')
σ_{p} surface stress due to external constraints or prestressing
τ point in time
$\vec{r}$ crack location
φ crack orientation

A crack can only grow or fail if it is exposed to tensile stress, i. e. if $\sigma_{\mathrm{n}}(t, \vec{r}, \varphi) > 0$. By considering negative σ_{n} as $\sigma_{\mathrm{n}} = 0$ in crack growth calculations, the effect of residual stresses can be accounted for in a simple and consistent way and the same design equations or algorithms can be used for all glass types.

From Equation (3.10) follows that the fracture strength of heat-treated glass is the sum of the absolute value of the residual (compressive) surface stress and of the strength of the glass itself, called *inherent strength* henceforth. Only the latter is influenced by sub-critical crack growth and depends, therefore, on time and environmental conditions (cf. Section 3.2). The residual stress is constant.

If a heat-treated glass element is designed such that (T is the design life)

$$\max_{\tau\in[0,T]} \sigma_{E,\mathrm{n}} \leq -(\sigma_{\mathrm{r,n}} + \sigma_{\mathrm{p,n}}) \ \forall(\vec{r}, \varphi)\,, \tag{3.11}$$

no surface decompression occurs, i. e. no surface crack is ever exposed to tensile stress during the entire service life. Such an element does not show any size-dependent, time-dependent or environment-dependent effects.

3.3.3 Inert strength

From Equation (3.8) and Equation (3.9), we can determine the stress intensity required for a crack to extend immediately and cause failure:

$$Y \cdot \sigma_{\mathrm{n}} \cdot \sqrt{\pi \cdot a} \geq K_{\mathrm{Ic}} \tag{3.12}$$

With this, the stress causing failure of a crack of depth a, the *critical stress* σ_{c}, is

$$\sigma_{\mathrm{c}}(t) = \frac{K_{\mathrm{Ic}}}{Y \cdot \sqrt{\pi \cdot a(t)}} \tag{3.13}$$

while the depth of a crack failing at the stress σ_{n}, the *critical crack depth* a_{c}, is

$$a_{\mathrm{c}}(t) = \left(\frac{K_{\mathrm{Ic}}}{\sigma_{\mathrm{n}}(t) \cdot Y\sqrt{\pi}}\right)^2 \tag{3.14}$$

Both the stress σ_{n} and the crack depth a are time-dependent. Therefore, σ_{c} and a_{c} are also time dependent. The critical stress represents the resistance of a crack to instantaneous failure (i. e. failure that is not triggered by sub-critical crack growth) and is therefore called *inert strength* henceforth. It is plotted in *Figure 3.5* as a function of the crack depth using typical parameters for a long, macroscopic surface crack of small depth in a glass plate.

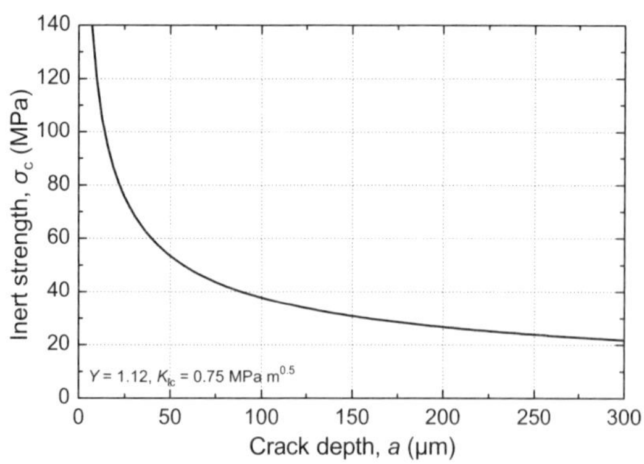

Figure 3.5:
Strength of a single crack at inert condition as a function of its depth.

3.3.4 Lifetime of a single flaw

Assuming the ordinary differential equation of crack growth (cf. Equation (3.3))

$$v = \mathrm{d}a/\mathrm{d}t = v_0 \left(K_\mathrm{I}/K_\mathrm{Ic}\right)^n \tag{3.15}$$

to be valid over the full range of K_I (which means neglecting the crack growth threshold), using the stress intensity factor from Equation (3.8) and assuming n to be constant, variable separation yields

$$\int_{a_\mathrm{i}}^{a(t)} a^{-\frac{n}{2}}\,\mathrm{d}a = \int_0^t v_0 \cdot K_\mathrm{Ic}^{-n} \cdot \left(Y\sqrt{\pi}\right)^n \cdot \sigma_\mathrm{n}^n(\tau)\,\mathrm{d}\tau \tag{3.16}$$

with a_i being the initial crack depth ($a_\mathrm{i} = a(t=0)$). The time-dependent size of a single crack exposed to the crack opening stress history $\sigma(t)$ is thus:

$$a(t) = \left[a_\mathrm{i}^{\frac{2-n}{2}} + \frac{2-n}{2} \cdot v_0 \cdot K_\mathrm{Ic}^{-n} \cdot \left(Y\sqrt{\pi}\right)^n \cdot \int_0^t \sigma_\mathrm{n}^n(\tau)\,\mathrm{d}\tau\right]^{\frac{2}{2-n}} \tag{3.17}$$

Variable separation and integration over the time interval $[0, T]$ and the corresponding crack depths $[a_\mathrm{i}, a]$ gives the following basic relationship:

$$\int_0^T \sigma_\mathrm{n}^n(\tau)\,\mathrm{d}\tau = \frac{2}{(n-2) \cdot v_0 \cdot K_\mathrm{Ic}^{-n} \cdot (Y\sqrt{\pi})^n \cdot a_\mathrm{i}^{(n-2)/2}} \left[1 - \left(\frac{a_\mathrm{i}}{a}\right)^{(n-2)/2}\right] \tag{3.18}$$

The crack depth at failure a_f is the critical crack depth (Equation (3.14)) for the failure stress $\sigma(t_\mathrm{f})$

$$a_\mathrm{f} = \left(\frac{K_\mathrm{Ic}}{\sigma_\mathrm{n}(t_\mathrm{f}) \cdot Y\sqrt{\pi}}\right)^2 \tag{3.19}$$

with t_f being the time to failure or lifetime of the crack in question. This can now be inserted into Equation (3.18). As n is large (≈ 16), the expression in square brackets in Equation (3.18) approaches 1 for long lifetimes with $a_\mathrm{f} \gg a_\mathrm{i}$. Thus the following simplified expression may be obtained:

$$\int_0^{t_\mathrm{f}} \sigma_\mathrm{n}^n(\tau)\,\mathrm{d}\tau = \frac{2}{(n-2) \cdot v_0 \cdot K_\mathrm{Ic}^{-n} \cdot (Y\sqrt{\pi})^n \cdot a_\mathrm{i}^{(n-2)/2}} \tag{3.20}$$

Given a stress history, this widely used relationship enables the calculation of the lifetime of a crack given its initial depth or the allowable initial crack depth given its required lifetime. The left side of Equation (3.20) is called *risk integral* or *'Brown's integral'*, because it was first used by Brown [47] to characterize damage accumulation in glass.

While Equation (3.20) is very convenient, it suffers from its limit of validity. If crack velocity is slow and/or the loading time is very short (near-inert conditions, see Section 6.4), a crack's strength as obtained from this equation converges on infinity. This of course, makes no sense. A crack's strength cannot be higher than the inert strength. The reason for the problem is that the crack depth at failure is not much bigger than the initial crack depth in the aforementioned conditions. In fact, in perfectly inert conditions, both depths are identical. The assumption of $a_\mathrm{f} \gg a_\mathrm{i}$ used to obtain Equation (3.20) is therefore *not* valid in such conditions.

From Equation (3.17) and Equation (3.19), a formulation of general validity can be obtained:

$$\tilde{a}_c(\tau) = \left[\left(\frac{\sigma_\mathrm{n}(\tau) \cdot Y\sqrt{\pi}}{K_\mathrm{Ic}} \right)^{n-2} + \frac{n-2}{2} \cdot v_0 \cdot K_\mathrm{Ic}^{-n} \cdot (Y\sqrt{\pi})^n \cdot \int_0^\tau \sigma_\mathrm{n}^n(\tilde{\tau})\,\mathrm{d}\tilde{\tau} \right]^{\frac{2}{2-n}} \tag{3.21}$$

The crack depth $\tilde{a}_c(\tau)$ is the initial depth of a crack that fails at the point in time τ when exposed to the crack-opening stress history $\sigma_\mathrm{n}(\tau)$. The choice of the symbol will become clear in Section 3.3.5. The disadvantage of Equation (3.21) is that it depends not only on the risk integral but also on the momentary stress $\sigma_\mathrm{n}(\tau)$. While the risk integral is monotonously increasing, in general the momentary stress is not. Therefore, the minimum initial crack depth $\min(\tilde{a}_c(\tau))$, which is relevant for design, does not necessarily occur at the end of the stress history ($\tau = T$) but may occur at any $\tau \in [0, T]$. A crack does not fail if $a_\mathrm{i} < \min_{\tau\in[0,T]} \tilde{a}_c(\tau)$.

Figure 3.6 quantitatively illustrates the behaviour of a surface crack using Equation (3.21). The curves show the constant stress that causes failure as a function of the loading time and for different initial crack depths. The figure is plotted for $v_0 = 6\,\mathrm{mm/s}$, which is the conservative assumption for structural design purposes given in Section 3.2. It can be seen that the strength of cracks is strongly time-dependent. Furthermore, the long-term strength of cracks with an initial depth in the order of 100 µm or more is low.

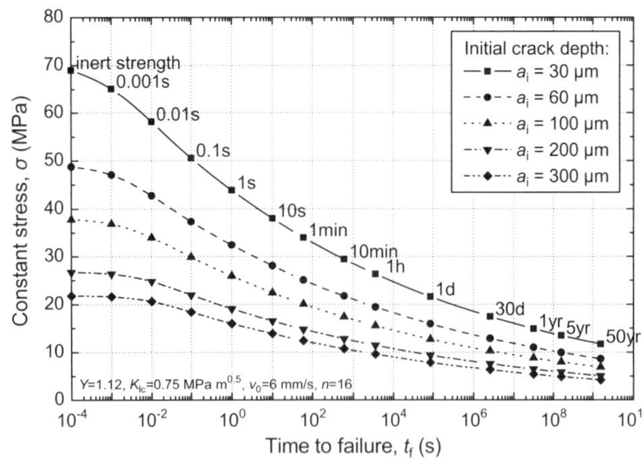

Figure 3.6:
Strength of a surface crack as a function of the loading time and the initial crack depth. The figure is based on a conservative assumption with regard to the crack growth behaviour, which is suitable for structural design.

Equivalent static stress and resistance

The simplified expression in Equation (3.20) is generally sufficient for structural design (but not necessarily for the interpretation of test results, see Chapter 6). This equation means that, if n is constant, two stress histories $\sigma_{(1)}(\tau)\ \tau \in [0, t_1]$ and $\sigma_{(2)}(\tau)\ \tau \in [0, t_2]$

cause the same crack growth if $\int_0^{t_1}\sigma_{(1)}^n(\tau)\,\mathrm{d}\tau = \int_0^{t_2}\sigma_{(2)}^n(\tau)\,\mathrm{d}\tau$. One can, therefore, define a *t_0-equivalent stress* σ_{t_0} as follows:[14]

$$\sigma_{t_0} = \left(\frac{1}{t_0}\int_0^T \sigma^n(\tau)\,\mathrm{d}\tau\right)^{1/n} \approx \left(\frac{1}{t_0}\sum_{j=1}^{J}\left[\sigma_{t_j}^n \cdot t_j\right]\right)^{1/n} \tag{3.22}$$

This equivalent stress is the stress that would, when applied during the *reference time period* t_0, cause the same amount of crack growth as the original stress history $\sigma(\tau)$. The right side of Equation (3.22) caters for discrete stress histories consisting of J time periods of duration t_j ($T = \sum t_j$) and constant stress σ_{t_j}. The same approach can be used for a crack's resistance by defining the *t_0-equivalent resistance*:

$$\sigma_{\mathrm{R},t_0} = \left(\frac{1}{t_0}\frac{2}{(n-2)\cdot v_0 \cdot K_{\mathrm{Ic}}^{-n}\cdot (Y\sqrt{\pi})^n \cdot a_{\mathrm{i}}^{(n-2)/2}}\right)^{1/n} \tag{3.23}$$

This is the static stress that a crack can resist for a reference time period t_0 (usually $t_0 = 1\,\mathrm{s}$, $3\,\mathrm{s}$ or $60\,\mathrm{s}$). It is independent of the applied load and completely characterizes the load resistance of a given crack (or an element whose load capacity is governed by this crack) for given environmental conditions (v_0, n), initial crack depth (a_i) and crack geometry (Y). A structural safety verification based on this approach entails ensuring that:

$$\sigma_{t_0} \leq \sigma_{\mathrm{R},t_0} \tag{3.24}$$

The relationship between lifetimes and applied constant stresses of two identical cracks (a_i, Y) in identical conditions (v_0, n, K_Ic) follows directly from Equation (3.22):

$$\frac{\sigma_2}{\sigma_1} = \left(\frac{t_1}{t_2}\right)^{1/n} \quad \text{or} \quad \frac{t_1}{t_2} = \left(\frac{\sigma_2}{\sigma_1}\right)^n \tag{3.25}$$

Inserting $\sigma(\tau) = \dot{\sigma}_{\mathrm{const}} \cdot \tau$ into Equation (3.22), the relationship between the lifetime of two identical cracks (a_i, Y) in identical conditions (v_0, n, K_Ic) loaded at constant stress rates $\dot{\sigma}_1$ and $\dot{\sigma}_2$ is obtained:

$$\frac{t_1}{t_2} = \left(\frac{\dot{\sigma}_2}{\dot{\sigma}_1}\right)^{\frac{n}{n+1}} \tag{3.26}$$

Since these equations are independent of v_0, they can be used to determine n. Plotting the failure stress as a function of the stress rate on logarithmic scales results in a slope of $1/(n+1)$. This allows the parameter n to be determined from experiments with variable stress rate. It should not be overlooked, however, that while the equations are independent of v_0, their validity is confined to cases in which flaws and conditions, *including* v_0, are identical during all tests. Since v_0 can be strongly stress rate dependent (cf. Section 3.2), this method should be used with caution.

[14]For hand calculations with simple stress histories, it is useful to express the T-equivalent stress σ_T in terms of a chosen, characteristic value σ_ch and a shape coefficient g with $\sigma_T = g^{1/n}\sigma_\mathrm{ch}$ and $g = T^{-1}\int_0^T (\sigma(\tau)/\sigma_\mathrm{ch})^n\,\mathrm{d}\tau$. For a constant stress σ_const, the coefficients are $\sigma_\mathrm{ch} = \sigma_\mathrm{const}$ and $g = 1$. For a constant stress rate ($\sigma(\tau) = \dot{\sigma}_\mathrm{const}\cdot\tau$), it is $g = 1/(n+1)$ and $\sigma_\mathrm{ch} = \max(\sigma(\tau))$. Values for other common stress histories are provided in [254].

3.3.5 Lifetime of a glass element with a random surface flaw population

Starting point and hypotheses

Section 3.3.4 discussed the lifetime of a single crack. For several cases of practical relevance, this is an appropriate way of modelling the surface condition of structural glass elements. For others, however, it is not (see Chapter 6). Let us consider a simple yet common case: as-received glass, i. e. glass as it is delivered to the client. As stated in Section 1.2.2, the surface contains a large number of mechanical flaws of varying severity, which are not necessarily visible to the naked eye. This surface condition can be represented much better by a statistical approach, namely, a *random surface flaw population* (RSFP). The mathematical relations from Section 3.3.4 need, therefore, to be extended to describe glass elements in which resistance is governed by such an RSFP. Only a very succinct derivation is presented in the following. For more information, the interested reader should refer to the detailed derivation provided by Haldimann [189].

In addition to the hypotheses used to predict the lifetime of a single flaw in Section 3.3.4, a few additional hypotheses are required for the present case:

1. The material contains a large number of natural flaws of variable depth.
2. The crack depth is a random variable[15] that can be represented by a statistical distribution.
3. The individual flaws do not influence each other.[16]
4. A glass element fails when the first flaw fails.
5. All crack locations and orientations have the same probability of occurrence.
6. Pure mode I crack propagation and failure represents the actual multimodal behaviour with sufficient accuracy.

For an in-depth assessment of the hypotheses used in Section 3.3.4 and the present section, see [189, Chapter 5].

In addition to linear elastic fracture mechanics, the present section makes use of the fundamental work in the fields of theory of probability and strength of materials, including [29, 30, 164, 336, 337].

Constant, uniform, and uniaxial stress

In order to make the derivation as clear and understandable as possible, two more very restrictive assumptions are made to start with, but will be dropped in the course of the generalization:

1. The orientation of all flaws is identical and perpendicular to the homogeneous tensile stress σ.
2. There is no sub-critical crack growth.

[15] One could also consider the *strength* of the flaws as the basic random variable. The choice is irrelevant because both quantities can be expressed in terms of each other using linear elastic fracture mechanics.

[16] This assumption is conservative. The presence of cracks modifies the stress field within the material. If the length of a surface crack is similar to, or longer than, the distance separating the cracks, it induces a shielding of the stress at the neighbouring crack tip. This effect can reduce crack growth and increase lifetime [25].

With a constant stress and no sub-critical crack growth, the crack depth a and the critical crack depth a_c (cf. Section 3.3.3) are both constant. The failure probability of a crack is simply the probability that its random size a is larger than the critical crack depth a_c:

$$P^{(1)}_{\mathrm{f,inert}}(a) = \mathscr{P}\left(a \geq a_c\right) = \int_{a_c}^{\infty} f_a(a)\,\mathrm{d}a = 1 - F_a\left(a_c\right) \tag{3.27}$$

F_a is the cumulative distribution function (CDF), f_a is the probability density function (PDF) of the crack depth. The strength distribution mainly depends on the distribution of the larger flaws. Assuming that the mean number of flaws is large, the mathematical theory of extreme values applies and shows that the asymptotic behaviour of the crack depth distribution can be described accurately by a power law. The probability density function (PDF) of the crack depth a is thus ($\propto$ means 'proportional to', r is a parameter):

$$f_a(a) \propto a^{-r} \tag{3.28}$$

The CDF of the crack depth $F_a = \int f_a$ is a *Pareto distribution*:

$$F_a(a) = \begin{cases} 0 & \text{for } a \leq a_0 \\ 1 - (a_0/a)^{r-1} & \text{for } a > a_0 \end{cases} \tag{3.29}$$

For normalization reasons of the CDF ($F_a = 1$ for $a \to \infty$), a lower limit a_0 for the crack depth a has to be introduced. Since very small cracks are irrelevant for failure, the actual value of a_0 is unimportant. Equation (3.29) sufficiently describes the crack distribution in the range of relevant crack depths.

An element fails if *any* of the flaws fail, or survives if *all* flaws survive. The *survival probability* of a glass element is, therefore, the product of the survival probabilities of all flaws. By considerable rearrangement, but without introducing additional simplifying assumptions, the *inert failure probability* $P_{\mathrm{f,inert}}$ of a glass element can be found:

$$P_{\mathrm{f,inert}}(\sigma) = 1 - \exp\left\{-\frac{A}{A_0}\left(\frac{\sigma}{\theta_0}\right)^{m_0}\right\} \tag{3.30}$$

$$\theta_0 = \frac{K_{\mathrm{Ic}}}{M_0^{1/m_0} \cdot Y\sqrt{\pi} \cdot \sqrt{a_0}} \qquad m_0 = 2(r-1) \tag{3.31}$$

The parameters θ_0 and m_0 *solely* depend on the surface flaw population and are therefore *true material parameters*. They can be determined from tests (see Section 6.4.2). High values of m_0 represent a narrow distribution of the crack depths, and therefore, of inert strengths. High values of θ_0 represent a high mean and wide distribution of inert strengths. Equation (3.30) can be rearranged to take on the form of a *two-parameter Weibull distribution* with scale parameter θ_{inert} and shape parameter m_0:

$$P_{\mathrm{f,inert}}(\sigma) = 1 - \exp\left\{-\left(\frac{\sigma}{\theta_{\mathrm{inert}}}\right)^{m_0}\right\} \tag{3.32}$$

$$\theta_{\mathrm{inert}} = \theta_0 \left(\frac{A}{A_0}\right)^{-1/m_0} \tag{3.33}$$

For two elements with surface areas A_1 and A_2 exposed to tensile stress, it is:

$$\frac{\theta_{\text{inert},A_1}}{\theta_{\text{inert},A_2}} = \left(\frac{A_2}{A_1}\right)^{1/m_0} \tag{3.34}$$

This ratio is commonly referred to as *size effect*. The influence of the size effect is small for high values of m_0 (small scatter of strength) and large for small values of m_0 (large scatter).

From Equation (3.32), the *reference inert strength* $f_{0,\text{inert}}(P_{\text{f,t}}, \theta_0, m_0)$, a quantity that will be useful when discussing structural design, can be defined:

$$f_{0,\text{inert}}(P_{\text{f,t}}, \theta_0, m_0) = \theta_0 \left[-\ln(1-P_{\text{f,t}})\right]^{1/m_0} \tag{3.35}$$

The physical meaning of $f_{0,\text{inert}}$ is as follows: A glass surface of area $A_0 = 1\,\text{m}^2$ fails with probability $P_{\text{f,t}}$ when exposed to a uniformly distributed crack opening surface stress $f_{0,\text{inert}}$ at inert conditions (see Section 6.4). The reference inert strength depends on the target failure probability and the glass surface condition *only*. It does *not* depend on the glass type (because it refers to the crack opening stress) or the crack velocity parameters (because it refers to inert conditions).

Extension to non-uniform, biaxial stress fields

In a *non-uniform stress field*, the stress, σ, depends on the point on the surface $\vec{r} = (x, y)$. Equation (3.30) can be extended accordingly by integrating over infinitesimal surface elements, dA, of constant stress. To be able to account for random crack orientation in a *biaxial stress field*, a multimodal failure criterion needs to be chosen. Published research suggests that the simplest failure criterion, pure mode I fracture, gives the best agreement with experimental results [189]. This simply means that a crack fails due to unstable crack propagation if $K_\text{I} > K_\text{Ic}$, that the mode I-equivalent stress is equal to the stress component perpendicular to the crack and that the mode I geometry factor Y can be used. With this notation, Equation (3.14) remains valid even for biaxial stress fields.

For a surface crack of orientation φ in plane stress state ($\sigma_z = \tau_{zx} = \tau_{zy} = 0$), the stress component normal to the crack is

$$\sigma_\text{n} = \sigma_1 \cos^2\varphi + \sigma_2 \sin^2\varphi \tag{3.36}$$

where σ_1 and σ_2 are the major and minor in-plane principal stresses ($\sigma_1 \geq \sigma_2$) and φ is the crack orientation (a crack with $\varphi = 0 \Rightarrow$ is parallel to the direction of σ_1).

Since glass is a homogeneous, isotropic material, it may be assumed that all crack locations $\vec{r} = (x, y)$ and crack orientations φ have the same probability of occurrence as long as no directional scratching is introduced. The probability density functions for a crack's location and orientation are, thus, both *uniform distributions* ($f_A(\vec{r}) = 1/A$ and $f_\varphi(\varphi) = 1/\pi$), such that the probability of finding a crack of orientation φ within the infinitesimally small surface area dA at the point $\vec{r}$ on the surface is $P_{\varphi,\vec{r}} = 1/A\,\text{d}A \cdot 1/\pi\,\text{d}\varphi$. With this, the inert failure probability of an element with a random number of randomly distributed and randomly oriented surface cracks is [189]:

$$P_{\text{f,inert}} = 1 - \exp\left\{-\frac{1}{A_0}\int_A \frac{2}{\pi}\int_{\varphi=0}^{\pi/2} \left(\frac{\sigma_\text{n}(\vec{r},\varphi)}{\theta_0}\right)^{m_0} \text{d}A\,\text{d}\varphi\right\} \tag{3.37}$$

Extension to time-dependent loading

In the absence of sub-critical crack growth, a crack fails at the point in time t if its depth a is larger than the momentary critical crack depth $a_c(t)$ (cf. Section 3.3.3). The probability that an individual crack will fail at time t is thus:

$$P^{(1)}_{\mathrm{f,inert}}(t) = \mathscr{P}\left(\exists\tau\in[0,t]: a \geq a_c(\tau)\right) = \mathscr{P}\left(a \geq \min_{\tau\in[0,t]} a_c(\tau)\right) \tag{3.38}$$

With this and the cumulative distribution function of the crack depth from Equation (3.29), the failure probability for time-dependent loading, but still without sub-critical crack growth, can be found [189]:

$$P_{\mathrm{f,inert}}(t) = 1 - \exp\left\{-\frac{1}{A_0}\int_A \frac{2}{\pi}\int_{\varphi=0}^{\pi/2}\left(\frac{\max\limits_{\tau\in[0,t]} \sigma_{\mathrm{n}}(\tau,\vec{r},\varphi)}{\theta_0}\right)^{m_0} \mathrm{d}A\,\mathrm{d}\varphi\right\} \tag{3.39}$$

Extension to account for sub-critical crack growth

Sub-critical crack growth (cf. Section 3.2) makes the surface flaw population time-dependent. Compared to Equation (3.38), not only the *critical*, but also the *momentary*, crack depth is now time-dependent:

$$P^{(1)}_{\mathrm{f}}(t) = \mathscr{P}\left(\exists\tau\in[0,t]: a(\tau) \geq a_c(\tau)\right) \tag{3.40}$$

The criterion for the initial crack depth given in Equation (3.21) enables Equation (3.40) to be expressed as:

$$P^{(1)}_{\mathrm{f}}(t) = \mathscr{P}\left(\exists\tau\in[0,t]: a_{\mathrm{i}} \geq \tilde{a}_c(\tau)\right) = \mathscr{P}\left(a_{\mathrm{i}} \geq \min_{\tau\in[0,t]} \tilde{a}_c(\tau)\right) \tag{3.41}$$

This means that instead of a criterion for the *momentary* crack depth $a(\tau)$, there is now a criterion for the *initial* crack depth a_{i}. One can, therefore, proceed in the same way as above, where the crack depth was time-independent and thus always equal to a_{i}. Through considerable rearrangement, but without introducing additional simplifying assumptions, an expression for the time-dependent failure probability of a general glass element that takes sub-critical crack growth, non-homogeneous time-variant biaxial stress fields, arbitrary geometry and arbitrary stress histories into account, can be found [189]:

$$P_{\mathrm{f}}(t) = 1 - \exp\left\{-\frac{1}{A_0}\int\limits_A \frac{2}{\pi}\int\limits_{\varphi=0}^{\pi/2}\left[\max_{\tau\in[0,t]}\left\{\left(\left(\frac{\sigma_{\mathrm{n}}(\tau,\vec{r},\varphi)}{\theta_0}\right)^{n-2} + \frac{1}{U\cdot\theta_0^{n-2}}\int\limits_0^{\tau}\sigma_{\mathrm{n}}^{n}(\tilde{\tau},\vec{r},\varphi)\,\mathrm{d}\tilde{\tau}\right)^{\frac{1}{n-2}}\right\}\right]^{m_0}\mathrm{d}A\,\mathrm{d}\varphi\right\} \tag{3.42}$$

A_0	unit surface area ($A_0 = 1\,\mathrm{m}^2$)
A	surface area of the glass element (both faces)
t	point in time
$\sigma_\mathrm{n}(\tau, \vec{r}, \varphi)$	in-plane surface stress component normal to a crack of orientation φ at the point $\vec{r}(x, y)$ on the surface and at time τ (cf. Equation (3.36))
θ_0, m_0	surface condition parameters as defined in Equation (3.31), to be determined from experiments; $[\theta_0] = \text{stress}$, $[m_0] = \text{none}$
U	combined coefficient containing parameters related to fracture mechanics and sub-critical crack growth; $U = 2K_\mathrm{Ic}^2 / \left[(n-2) \cdot v_0 \cdot Y^2 \pi\right]$; $[U] = \text{stress}^2 \cdot \text{time}$
K_Ic	fracture toughness (cf. Section 3.3.1)
v_0, n	crack velocity parameters (cf. Section 3.2.1)
Y	geometry factor (cf. Section 3.3.1)

For an in-depth discussion of Equation (3.42), the interested reader should refer to [189, in particular Chapter 5].

Simplification for structural design

The following simplifications are appropriate for the vast majority of common structural glass design tasks (see [189, Chapter 5]):

- Calculating the failure probability on the basis of the risk integral is an approximation of sufficient accuracy.
- The crack growth threshold can be neglected.
- An equibiaxial stress field may be assumed.

These assumptions enable the model from Section 3.3.5 to be simplified substantially. Using the t_0-second major principal stress (cf. Equation (3.22))

$$\sigma_{1,t_0}(t, \vec{r}) = \left(\frac{1}{t_0} \int_0^t \sigma_1^n(\tau, \vec{r})\, \mathrm{d}\tau \right)^{1/n}, \tag{3.43}$$

Equation (3.42) simplifies to

$$P_\mathrm{f}(t) = 1 - \exp\left\{ -\frac{1}{A_0} \left(\frac{t_0}{U \cdot \theta_0^{n-2}} \right)^{\frac{m_0}{n-2}} \int_A \left[\sigma_{1,t_0}(t, \vec{r})\right]^{\frac{n\,m_0}{n-2}} \mathrm{d}A \right\}. \tag{3.44}$$

However, this is just a model and not a design equation. For design, the failure probability is not a result but a target value. This means that the target failure probability needs to be introduced as an additional parameter. Furthermore, the standard verification format that engineers are used to involves the comparison of a resistance term to an action term. Equation (3.44) needs to be reformulated accordingly.

The first step is to define a uniformly distributed stress $\sigma_{1\mathrm{eq},t_0}$ that would have the same effect as the actual stress distribution, namely $\int_A \sigma_{1,t_0}^{\bar{m}}\, \mathrm{d}A = A \cdot \sigma_{1\mathrm{eq},t_0}^{\bar{m}}$. Using the combined parameter $\bar{m} = n\,m_0/(n-2)$, the *equivalent uniformly distributed stress* is:

$$\sigma_{1\mathrm{eq},t_0} = \left(\frac{1}{A} \int_A \sigma_{1,t_0}^{m}\, \mathrm{d}A \right)^{1/\bar{m}} \tag{3.45}$$

Secondly, this stress value can be further standardized by defining the equivalent stress that would have the same effect when acting on the unit surface area $A_0 = 1\,\text{m}^2$ instead of A. This yields the *equivalent t_0-second uniform stress on the unit surface area (in short: equivalent reference stress)*:

$$\bar{\sigma} = \sigma_{1\text{eq},t_0,A_0} = \left(\frac{A}{A_0}\right)^{1/\bar{m}} \sigma_{1\text{eq},t_0} = \left(\frac{1}{A_0}\int_A \sigma_{1,t_0}^{\bar{m}}\,\text{d}A\right)^{1/\bar{m}} \tag{3.46}$$

Inserting this into Equation (3.44) and introducing one more combined parameter ($\bar{k}$) yields:

$$P_\text{f}(t) = 1 - \exp\left\{-\bar{k}\bar{\sigma}^{\bar{m}}\right\} \tag{3.47}$$

$$\bar{m} = \frac{n\,m_0}{(n-2)} \qquad \bar{k} = \left(\frac{t_0}{U\cdot\theta_0^{n-2}}\right)^{\frac{m_0}{n-2}} \tag{3.48}$$

Provided that the stress history of all sub-surfaces is known, $\bar{\sigma}$ can be evaluated for any condition. Rearrangement yields a standard failure criterion:

$$\bar{\sigma} < f_0(P_\text{f,t}) \tag{3.49}$$

$$f_0(P_\text{f,t}) = \left[-\ln(1-P_\text{f,t})\right]^{1/m}\cdot\left(\frac{t_0}{U\cdot\theta_0^{n-2}}\right)^{-1/n} = f_{0,\text{inert}}^{(n-2)/n}\cdot\left(\frac{U}{t_0}\right)^{1/n} \tag{3.50}$$

The reference inert strength $f_{0,\text{inert}}$ is defined in Equation (3.35) and U is defined in Equation (3.41). The resistance term $f_0(P_\text{f,t})$, called *reference ambient strength* hereafter, is a function of the *target failure probability* $P_\text{f,t}$ and has the following physical meaning: The failure probability of a glass element with surface area $A_0 = 1\,\text{m}^2$ that is exposed to a uniformly distributed crack opening surface stress f_0 for $t_0 = 1\,\text{s}$ at ambient conditions is $P_\text{f,t}$. It is important to be aware of the parameter dependencies:

- The equivalent reference stress $\bar{\sigma}$ is a function of the loading history (intensity and shape), the residual stress σ_r, the element surface area A, the exponential crack velocity parameter n, and the surface condition parameter m_0.
- The reference inert strength $f_{0,\text{inert}}$ is a function of the target failure probability $P_\text{f,t}$ and of the surface condition parameters θ_0 and m_0 only. The reference ambient strength f_0 depends additionally on the crack velocity parameters v_0 and n. In contrast to common measures for glass resistance (cf. Chapter 4), however, it is independent of the loading history, the surface area A, and the residual stress σ_r.

The lifetime prediction model from Equation (3.42) was already simplified considerably above. The quantification of the equivalent reference stress (Equation (3.46)), however, still requires a transient finite element analysis. This is unproblematic for research activities, for simple loading histories, and for the interpretation of experimental data. For application in practice, however, such analyses are often too complex and time-consuming . It is, therefore, pertinent to discuss how and under what conditions they can be avoided.

A time-dependent non-uniform stress field $\sigma(\tau,\vec{r})$ can be expressed in terms of a *representative stress* $\breve{\sigma}(\tau)$ at one point on the surface and a *dimensionless stress distribution function* $c(\tau,\vec{r})$:

$$\sigma(\tau,\vec{r}) = \breve{\sigma}(\tau)\cdot\left(\frac{\sigma(\tau,\vec{r})}{\breve{\sigma}(\tau)}\right) = \breve{\sigma}(\tau)\cdot c(\tau,\vec{r}) \tag{3.51}$$

The maximum stress on an element $\sigma_{\max}(\tau)$ is generally a sensible choice for the representative stress $\breve{\sigma}(\tau)$. It is important to bear in mind that $\sigma(\tau,\vec{r})$ refers to the crack opening stress (cf. Section 3.3.2), such that $c(\tau,\vec{r}) = \sigma(\tau,\vec{r})/\breve{\sigma}(\tau)$ is equal to zero in the compressed regions of the surface.

The dimensionless stress distribution function allows for the rearrangement of Equation (3.46) as follows:

$$\bar{\sigma} = \left(\frac{1}{A_0}\int_A\left[\left(\frac{1}{t_0}\cdot\int_0^t (\breve{\sigma}(\tau)\cdot c(\tau,\vec{r}))^n\,\mathrm{d}\tau\right)^{\bar{m}/n}\right]\mathrm{d}A\right)^{1/\bar{m}} \tag{3.52}$$

If $c(\tau,\vec{r})$ is independent of the load level represented by the time-dependent representative stress $\breve{\sigma}(\tau)$, and therefore, independent of the time τ, it can be isolated from the time-integral. This allows the time-integral and the area-integral to be separated such that the equivalent reference stress $\bar{\sigma}$ can be expressed as follows:

$$\bar{\sigma} = A_0^{-1/\bar{m}}\cdot\breve{\sigma}_{t_0}\cdot\bar{A}^{1/\bar{m}} \qquad \text{with} \qquad \bar{A} = \int_A c(\vec{r})^{\bar{m}}\,\mathrm{d}A \tag{3.53}$$

$\breve{\sigma}_{t_0}$ is the t_0-second *equivalent representative stress* (calculated from $\breve{\sigma}(\tau)$ using Equation (3.22)). The *equivalent area* $\bar{A}$ (also known as the *effective area*) is the surface area of a glass element that fails with the same probability, when exposed to the uniform representative stress $\breve{\sigma}$, as an element with surface area A when exposed to the non-uniform stress field $\sigma(\vec{r})$. $\bar{A}$ can be defined for ambient and inert conditions alike, with $\bar{m}$ being $n\,m_0/(n-2)$ and m_0 respectively.

The formulation in Equation (3.53) is of particular interest: It enables, for instance, convenient design aids to be created in order to avoid transient finite element analyses for common design tasks. In fact, all current glass design methods assume Equation (3.53) to be valid without declaring this assumption or discussing the conditions required for its validity. It should, however, be noted that Equation (3.53) is valid *only* if $\bar{A}$ and therefore the stress distribution function $c(\tau,\vec{r})$ are constant for all $\tau\in[0,t]$.

These findings allow the following conclusions to be drawn:

- **General conditions.** Geometric non-linearity (e. g. because plates undergo deformations larger than their thickness), residual stress (σ_r), external constraints (σ_p), and actions that vary not only in intensity but also in shape, make the dimensionless stress distribution function c and therefore the equivalent area $\bar{A}$, depend on the representative stress and therefore on time. Equation (3.53) is *not* valid in these conditions and there is no simple way to superimpose loads or to consider load duration effects. Therefore, Equation (3.46) must be solved.

- **Conditions in which no transient analysis is required.** Equation (3.53) allows transient analyses to be avoided if $\bar{A}$ and therefore the stress distribution function $c(\tau,\vec{r})$ are constant for *all* $\tau\in[0,t]$. This requires the following two conditions to be satisfied:

1. *The decompressed surface area remains constant.* This is the case if $\sigma(\tau,\vec{r}) > 0\,\forall(\tau,\vec{r})$ or if $\sigma(\tau,\vec{r}) \leq 0\,\forall(\tau,\vec{r})$. The first case occurs, for instance, on the tension face of annealed glass panes ($\sigma_r \approx 0$) exposed to a uniform lateral load. A typical example for the second case is a heat-treated glass element that is designed such that no surface decompression occurs (cf. Section 3.3.2).
2. *The major principal stress is proportional to the load at all points on the surface.* This requires in particular: a) linear elastic material behaviour, b) no (or negligible) geometrically non-linear behaviour, and c) variation of the applied load intensity only (but not of the load shape).

Under these conditions, $c(\tau,\vec{r})$ and $\bar{A}$ do *not* depend on the representative stress $\breve{\sigma}(\tau)$ and are therefore time-independent. They depend solely on the shape of the stress distribution within the element and on the element's size, both of which are in turn constant for the conditions at hand. Although there are many cases in which the conditions are not satisfied, current glass design methods implicitly assume that they are.

The common approach of applying the load duration effect found for a single crack (see Section 3.3.4) to the representative stress $\breve{\sigma}$ is valid under these special conditions:

$$\frac{\breve{\sigma}^{(2)}}{\breve{\sigma}^{(1)}} = \left(\frac{T_1}{T_2}\right)^{1/n} \tag{3.54}$$

From Equation (3.53), the ratio between the equivalent reference stresses $\bar{\sigma}$ of two load histories characterized by the t_0-second equivalent representative stresses $\breve{\sigma}_{t_0}$ can be derived:

$$\frac{\bar{\sigma}^{(1)}}{\bar{\sigma}^{(2)}} = \frac{\breve{\sigma}^{(1)}_{t_0}}{\breve{\sigma}^{(2)}_{t_0}} \tag{3.55}$$

Therefore, if the equivalent stress $\bar{\sigma}^{(1)}$ in a given loading situation of duration T_1 is known, the equivalent stress $\bar{\sigma}^{(2)}$ resulting from the same loading being applied for T_2 is:

$$\frac{\bar{\sigma}^{(2)}}{\bar{\sigma}^{(1)}} = \left(\frac{T_2}{T_1}\right)^{1/n} \tag{3.56}$$

As two elements with $\bar{\sigma}^{(1)} = \bar{\sigma}^{(2)}$ have equal probabilities of failure, Equation (3.53) provides a simple approach to design. The t_0-second resistance of specific elements in specific conditions can be provided, for instance, in design tables or graphs. A structural safety verification simply involves comparing this resistance to a t_0-second equivalent representative stress $\breve{\sigma}_{t_0}$ calculated from an arbitrary stress history. As the representative stress is proportional to the load ($\breve{\sigma}(t) = \varepsilon \cdot q(t)$) in conditions in which $\bar{A}$ is constant, this approach can even be extended to loads. Analogous to Equation (3.22), it is:

$$\mathbb{X}_{t_0} = \left(\frac{1}{t_0}\int_0^T \mathbb{X}^n(\tau)\,\mathrm{d}\tau\right)^{1/n} \approx \left(\frac{1}{t_0}\sum_{j=1}^{J}\left[\mathbb{X}^n_{t_j}\cdot t_j\right]\right)^{1/n} \quad \text{with } \mathbb{X} = \begin{cases} \breve{\sigma} & \text{for stresses} \\ q & \text{for loads} \end{cases} \tag{3.57}$$

➪ **Constant load and non-linear behaviour.** If a constant load is applied during T, $\bar{A}$ is also constant. It is

$$\breve{\sigma}_{t_0} = \breve{\sigma}\left(T/t_0\right)^{1/n} = \breve{\sigma}\bar{t} \tag{3.58}$$

and Equation (3.53) is applicable.

➪ **Comparing two periods of constant load on an element with non-linear behaviour.** For any two periods of duration T_1 and T_2 with constant representative stresses $\breve{\sigma}^{(1)}$ and $\breve{\sigma}^{(2)}$, it is:[17]

$$\frac{\breve{\sigma}^{(2)}}{\breve{\sigma}^{(1)}} = \left(\frac{T_1}{T_2}\right)^{1/n}\left(\frac{\bar{A}_1}{\bar{A}_2}\right)^{1/\bar{m}} \tag{3.59}$$

This means that the 'tensile strength ratio' , the ratio of the maximum allowable stress on a glass element for two periods of constant load, depends on the equivalent area $\bar{A}$, which is in general a function of the geometry, the stress level, and the shape of the load.

3.3.6 Discussion

While the lifetime prediction model described herein is more complex than traditional semi-empirical models, it offers significant advantages over these. A comprehensive and clear derivation enables the model and its hypotheses to be fully understood by its users. The model contains no simplifying hypotheses which would restrict its applicability to special cases. Its parameters have a clear physical meaning that is apparent to the engineer. They each include only one physical aspect and they do not depend on the experimental setup used for their determination. The condition of the glass surface can be modelled using either a single surface flaw or a random surface flaw population and the properties of these surface condition models are independent parameters that the user can modify. This is a major advantage, especially when hazard scenarios that involve surface damage must be analysed or when data from quality control measures or research are available. Finally, the material strength rightly converges on the inert strength for very short loading times or slow crack velocity.

Chapter 6 will discuss the use of this lifetime prediction model to overcome shortcomings of current design methods. This chapter also provides a table (*Table 6.1*) that clearly shows when to use which of the equations from the present chapter.

3.4 Dynamic fracture mechanics

The lifetime prediction model described in Section 3.3 provides a mathematical means for determining the fracture strength of glass by describing the condition up to and including the instant of failure. It does not, however, provide information on what happens after the fracture strength is exceeded. This information is vital for understanding post-failure phenomena, for example, in the diagnostic interpretation of glass failure, which is

[17] An equal probability of failure is obtained if the equivalent reference stresses are identical: $\bar{\sigma}^{(1)} \overset{!}{=} \bar{\sigma}^{(2)} \implies A_0^{-1/\bar{m}} \cdot \breve{\sigma}^{(1)}\bar{t}_1 \cdot \bar{A}_1^{1/\bar{m}} \overset{!}{=} A_0^{-1/\bar{m}} \cdot \breve{\sigma}^{(2)}\bar{t}_2 \cdot \bar{A}_2^{1/\bar{m}}$. Rearrangement and insertion of $\bar{t} = (T/t_0)^{1/n}$ yields Equation (3.59).

discussed in Section 8.2. In this case it is useful to be able to quantify the crack branching behaviour of glass. A universally agreed theoretical explanation of crack branching is still elusive. However, a number of possible explanations have been put forward. These are of formidable theoretical complexity and beyond the scope of this document. Interested readers may refer to specialized literature such as [171, 238]. Some of the simplified empirical formulations, however, are of practical interest. They are, therefore, presented hereunder. Their use for the diagnostic interpretation of glass failures is explained in Section 8.2.

If an unbalanced force acts on a crack, i. e. $K_\mathrm{I} \geq K_\mathrm{Ic}$, there is excess energy to drive the crack and the fracture becomes unstable. This is known as dynamic fracture and the equilibrium conditions of Griffith and Irwin no longer apply. Under these conditions, the crack propagates and accelerates very rapidly, typically between $1\,500 - 2\,500\,\mathrm{m/s}$ for soda lime silica glass. This phenomenon is therefore referred to as 'instantaneous' or 'catastrophic' failure. There are two ways in which a crack may become dynamic:

1. The crack reaches a point of instability because the applied stress or the crack depth causes the stress intensity factor K_I to exceed the critical value K_Ic. Since cracks grow under static loads, a dynamical state may be realized even under constant loading conditions. A running crack accelerates rapidly towards a terminal velocity governed by the speed of elastic waves.
2. The applied loading is subject to a rapid time variation, as in impact loading.

A general approach to the dynamic fracture problem was outlined by Mott [257] in an extension to the Griffith concept. He simply incorporated a term for the kinetic energy, U_K, into the expression for the total system energy (Equation (3.6)):

$$U = U_\mathrm{M} + U_\mathrm{S} + U_\mathrm{K} \tag{3.60}$$

The kinetic energy term accounts for the kinetic energy of the advancing crack. Mott was able to quantify U_K for various (though rather simple) geometries and loading conditions, such that the behaviour of a running crack can be predicted in terms of kinetic energy and crack velocity as a function of the crack depth. He had, however, to make very restrictive simplifying assumptions. He assumed, for instance, that a crack does not bifurcate or branch. Further issues that are not taken into account include the influence of stress waves that are reflected at the specimen boundaries and the fact that the microstructural processes in the crack tip area, which govern the crack growth behaviour, are not the same at high speeds as in quasi-static conditions.

Crack branching marks various stages of kinetic energy dissipation and is of major interest for fracture of soda lime silica glasses used in construction. The initial acceleration of the flaw starts on a relatively smooth surface known as the 'mirror zone'. As the flaw continues to accelerate, the higher stresses and greater energy released produce some form of micro-mechanical activity close to the crack tip, producing severe surface roughening that finally causes the crack to bifurcate or branch along its front. This is observed as an abrupt branching when the glass is viewed laterally. However, an elevation of the crack surface will reveal a progressive increase in the roughness of the fracture surface from 'mirror' to 'mist' to 'hackle' (*Figure 3.7*).

From the early 1950s, experiments were performed to ascertain the role of crack velocity in branching. Levengood [239] and Shand [308, 309] found empirically that

Figure 3.7:
Schematic representation of mirror, mist and hackle.

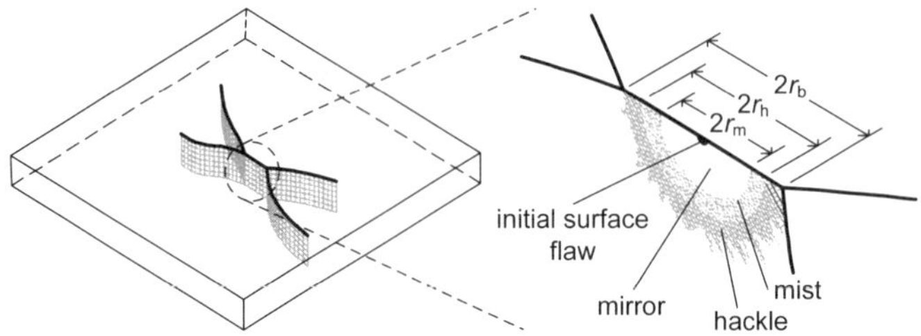

the fracture stress σ_f, i. e. the maximum principal tensile stress at the fracture origin, was approximately proportional to the reciprocal of the square root of the mirror radius (radius of the mirror/mist boundary) r_m:

$$\sigma_f = \alpha_m \cdot r_m^{-1/2} \tag{3.61}$$

Based on previous findings and further experimentation, Clark and Irwin [67] concluded that crack branching is primarily controlled by a critical value of the strain-energy release rate or stress intensity, rather than a crack-speed criterion. Though there is still much debate on the exact mechanism of crack branching, this interpretation is widely accepted today. Various experimental and theoretical efforts led to relationships of the same form as Equation (3.61) and although its theoretical background is still in dispute, this relationship found general acceptance because it is in reasonable agreement with experimental results. The relationship was found to be equally valid for the radius of the mist/hackle boundary r_h, and for one-half the crack length at macroscopic branching r_b (see [283] for a more detailed literature review), such that it can be rewritten in the more general form

$$\sigma_f = \alpha \cdot r^{-1/2} \tag{3.62}$$

where r is r_m, r_h, or r_b with the corresponding branching constants α_m, α_h, or α_b.

Duckworth, Shetty, Rosenfield and Siskos [92, 312] found that linear regression to experimental data always yielded finite intercepts and thus suggested a modification of Equation (3.62) to

$$\sigma_f - \sigma_{ar} = \alpha \cdot r^{-1/2} \tag{3.63}$$

where σ_{ar} was originally interpreted as being the residual compressive surface stress. An alternative explanation for σ_{ar} has since been put forward, it is therefore pertinent to term this quantity as apparent residual compressive surface stress. They furthermore concluded from their studies that the mirror constant is not influenced significantly by stress gradients in the specimen [92].

Reed and Bradt [286] determined the mirror constant α_m by analysing published failure data of unweathered and weathered window glass panels using Equation (3.62). The fact that their values ($\alpha_m = 1.92\,\mathrm{MPa\,m^{0.5}}$ for unweathered and $\alpha_m = 2.18\,\mathrm{MPa\,m^{0.5}}$ for weathered glass, assuming $\sigma_{ar} = 0$) were in close agreement to those determined in previous studies using small-scale laboratory testing showed that the relationship between the mirror radius and the failure stress may be extended to much larger structures such as windows panels.

Conway and Mecholsky [72] used the relationship

$$\sigma_{ar} r_m^{1/2} + \Psi_0 = \sigma_f r_m^{1/2} Y_F(\theta) \tag{3.64}$$

to predict the residual compressive surface stress σ_r, which they assumed to be equal to the apparent residual surface compression stress σ_{ar}, from the failure stress σ_f. ψ_0 is a material constant, $Y_F(\theta)$ is a crack-border correction factor. The angle θ indicates the point on the branching boundary ($\theta = 0°$: deepest point, $\theta = 90°$: point on the specimen surface). This means that while Equation (3.63) is valid only on the specimen surface, Equation (3.64) is, in principle, valid for all points along the branching boundary. This generalization remained, however, of limited practical interest because no published mirror/mist boundary data at other points than the specimen surface was available. Examination of Equation (3.64) indicates that for an ideally annealed glass plate ($\sigma_r = 0\,\text{MPa}$), a plot of $\sigma_f r_m^{1/2} Y_F(\theta)$ versus $r_m^{1/2}$ should yield a horizontal line having an ordinate of ψ_0. In the case of a tempered plate, a line with a positive slope that yields the magnitude of the residual stress and an intercept at ψ_0 should result. Conway and Mecholsky were able to show that the residual stress determined using this technique is indeed in relatively good agreement with direct residual stress measurements by optical techniques. The accuracy is, however, rather limited (tempered soda lime silica glass: 82 MPa from crack branching versus 96 MPa by birefringence measurement, annealed SLSG: 7 MPa versus 2 MPa), such that direct residual stress measurement remains preferable for diagnostic purposes.

Oakley [262] verified the accuracy of Equation (3.63) for the prediction of the macroscopic branch length $2r_b$ by testing a large series of 540 4 mm thick annealed float glass specimens containing only natural flaws in biaxial loading. Equation (3.63) fits well to his experimental results. Furthermore, the crack mirror constant $\alpha_b = 2.14\,\text{MPa}\,\text{m}^{0.5}$ and the apparent residual stress $\sigma_{ar,b} = 10.9\,\text{MPa}\,\text{m}^{0.5}$ determined from this data are similar to previously published results from both biaxial and uniaxial loading tests. This confirms the usefulness of the approach in diagnostic fracture analysis where the exact nature of the loading is generally uncertain. However, the apparent residual stress σ_{ar}, although similar to previous measurements, is clearly higher than the actual residual compressive surface stress σ_r of the samples. This casts doubt that σ_{ar} is an accurate measure of the residual stress. Oakley found from an analytical analysis that the slope of the curve (α_b) is insensitive to the plate thickness, but the intercept increases for thin plates. He therefore attributed the difference between apparent and actual residual stress to the effect of the finite plate thickness on the branching criterion when cracks are large.

Finally, all three branching constants α_m, α_h, and α_b as well as the corresponding apparent residual stresses σ_{ar} were determined in a recent study by Quinn [283]. He used experimental data from biaxial strength tests on annealed glass disks that were performed under a wide range of conditions, including different environments, stress rates, and both artificial and natural surface flaws. The following parameters were found: $\alpha_b = 2.28\,\text{MPa}\,\text{m}^{0.5}$, $\sigma_{ar,b} = 10.7\,\text{MPa}$; $\alpha_h = 2.11\,\text{MPa}\,\text{m}^{0.5}$, $\sigma_{ar,h} = 9.1\,\text{MPa}$ and $\alpha_m = 1.98\,\text{MPa}\,\text{m}^{0.5}$, $\sigma_{ar,m} = 9.6\,\text{MPa}$. Although the BK7 (a high quality optical bor-crown glass) used in these tests is not normally used in architectural applications, the study provides some additional insight. It is an experimental confirmation that the relationship between the fracture stress and the size of the measured fracture feature (r_m, r_h, or r_b) is constant over a wider range of conditions. This relationship is independent of the environment (dry nitrogen, air, water), the rate of applied stress, the surface condition, and the fracture stress. The fact that the parameters found are in good agreement with the values determined by Oakley on soda lime silica glass (cf. above) suggests that these conclusions are equally valid for soda lime silica glass and that the glass composition has a minor influence on the crack branching behaviour. Furthermore, Quinn suggests another alternative explanation for the difference between the apparent and the actual residual stress. He interprets the apparent residual stress he observed (about 10 MPa, cf. above) as a threshold stress below which crack branching does not occur.

The practical use of dynamic fracture mechanics for the diagnostic interpretation of glass failures is discussed in Section 8.2 and [265].

3.5 Laboratory testing procedures

In order to understand glass design, some knowledge about glass testing procedures is indispensable. Some of the most commonly used laboratory testing are, therefore, briefly outlined hereunder.

3.5.1 Testing procedures for crack velocity parameters

The following testing procedures are widely used to determine crack velocity parameters (see Section 3.2.1):

Direct measurement of the growth of large through-thickness cracks.

Particularly before measurements on indentation cracks (see below) became popular, this experimental approach was used to determine crack velocity parameters. The growth of a large through-thickness crack is directly measured as a function of the stress intensity factor. This can, for instance, be done optically or using sound waves. On one hand, this is a direct and relatively precise approach. On the other hand however, the behaviour of such large through-thickness cracks is not necessarily representative of the behaviour of the relatively small surface flaws that are relevant for structural design of glass elements. While Richter [289] (cf. above) could only measure crack velocities in the range of $10^{-5}\,\text{mm/s} \leq v \leq 10^{-2}\,\text{mm/s}$, which is clearly above the range that is relevant for structural glass design[18], modern technologies such as atomic force microscopy allow measurements within a wider range of $10^{-9}\,\text{mm/s} \leq v \leq 1\,\text{mm/s}$ [249].

Direct or indirect measurement of the growth of indentation flaws.

Since indentation flaws are relatively small surface flaws, they are more representative of the flaws governing failure of structural glass elements than long through-surface cracks. The advantage of indentation flaws over 'natural' surface flaws is that their fracture mechanics characteristics are well known, which is crucial if accurate crack velocity parameters are to be obtained. The growth of indentation flaws may either be observed directly or derived from ambient strength data.[19]

[18] At $10^{-5}\,\text{mm/s}$, a crack grows by 1 mm within 28 hours.

[19] For details on the procedure, see e. g. [179, 303, 306, 307].

3.5.2 Testing procedures for strength data

Static long-term tests

Static long-term tests with constant stress, also known as 'static fatigue tests', are usually performed using a four-point bending test setup. The testing procedure consists of applying a constant load and measuring the time to failure. The main advantage of such tests is their similarity with in-service conditions of structural glass elements that mainly carry dead loads. The disadvantage is that such tests are extremely time-consuming. If a specimen's surface condition or the stress corrosion behaviour differs only slightly from the assumptions used to design the test, the specimen may only fail after several years or not at all.

Dynamic fatigue tests

The term '*dynamic fatigue test*' is a generic term used for constant load rate testing, for constant stress rate testing, and for testing with cyclic loading. It is mostly performed using four-point bending (4PB) or coaxial double ring (CDR) test setups (also known as concentric ring-on-ring tests). *Figure 3.8* shows a schematic representation of the two test setups.

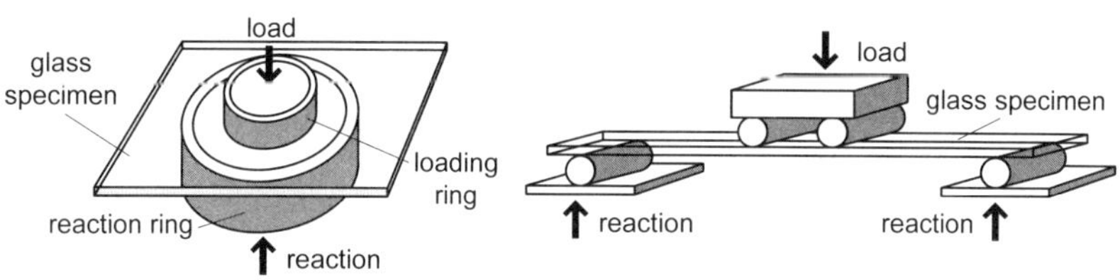

Figure 3.8: Schematic representation of coaxial double ring (left) and four-point bending (right) test setups.

In 4PB tests, the specimen is exposed to an approximately uniaxial stress field ($\sigma_1 \neq 0$, $\sigma_2 = 0$). In CDR tests, an equibiaxial stress field ($\sigma_1 = \sigma_2$) is obtained.[20]

Both test setups are simple and provide short time to failure even for specimens with small surface defects (e. g. as-received glass). The failure stress is a function of the stress rate. When plotting this relationship on logarithmic scales, a line with a slope of $1/(n+1)$ is obtained. If v_0 is constant, this allows for the determination of the crack velocity parameter n from tests at different stress rates.

In Europe, the testing procedure that is mostly used to obtain glass strength data is the *coaxial double ring test*. It is standardized in EN 1288-1:2000 [113] (fundamentals), EN 1288-2:2000 [114][21] (R400 test setup), and EN 1288-5:2000 [116] (R45 and R30 test setups). Details on the different setups are given in *Table 3.2*. Another common procedure, the *four-point bending test*, is standardized in EN 1288-3:2000 [115]. In all these tests, the stress rate to be used is $2 \pm 0.4\,\text{MPa/s}$.

[20] For detailed information on the CDR testing procedure, the interested reader should refer to seminal work on the subject such as [296] (basis for EN 1288-2:2000 [114], in German) or [318].

[21] DIN 52292-2:1986 [85], which was used for the majority of tests performed in the past, has been replaced by this standard. Apart from the suppression of the test setup R200, which was hardly ever used, it does not contain any relevant changes.

Table 3.2: Coaxial double ring test geometries in European standards.

Designation	Standard	Loading ring radius (mm)	Reaction ring radius (mm)	Tested area* (mm^2)	Specimen edge length (mm)
EN CDR R45	EN 1288-5 [116]	9	45	254	100 (±2)
EN CDR R400	EN 1288-2 [114]	300 ± 1	400 ± 1	240 000†	1 000 (±4)

* This is the surface area in uniform, equibiaxial tension = the area inside the loading ring (exception, see †).
† This is the value from the code. It does *not* correspond to the area within the load ring (282 743 mm^2).

Testing is mostly done on as-received glass specimens or specimens with artificially induced homogeneous surface damage. The data obtained represents a combination of the specimen's surface condition and the crack growth behaviour during the tests. Statistical analysis of the test results is generally done by fitting a two-parameter Weibull distribution [336, 337] to the experimental failure stress data:

$$P_{\mathrm{f}}(\sigma_{\mathrm{f},A}) = 1 - \exp\left[-\left(\frac{\sigma_{\mathrm{f},A}}{\theta_A}\right)^{\beta}\right] \tag{3.65}$$

$P_{\mathrm{f}}(\sigma_{\mathrm{f},A})$ is the cumulative probability of failure and $\sigma_{\mathrm{f},A}$ is the failure stress of specimens of which the surface area A is exposed to tensile stress. θ_A is the scale parameter (depends on A) and β the shape parameter of the Weibull distribution. Various methods for parameter estimation exist. The procedure standardized in EN 12603:2002 [106] was often used in the past.[22] It is based on point estimates and the median-rank based empirical failure probability given in Equation (C.6). For details on this approach as well as on alternative methods, see Section C.3.

For a general introduction to Weibull statistics, the interested reader should refer to a statistics book, e. g. [26, 256].

Tests for the glass failure prediction model

The underlying model of the US and Canadian Standards, the so-called glass failure prediction model (GFPM), does not use the above-mentioned testing procedures. The two interdependent surface flaw parameters $\tilde{m}$ and $\tilde{k}$ are determined by loading rectangular glass plates with uniform lateral load. The visually determined failure origin, the stress history at the failure origin and a rather complex iterative procedure are used to find the parameters (see Section 4.4.1). Only one crack velocity parameter, $n = 16$, is explicitly considered in the GFPM.

3.6 Quantitative considerations

3.6.1 Introduction

Now what is the actual strength of glass? In view of the complex material behaviour described in the preceding sections of this document, it is clear that there is no simple

[22] The older German national standard DIN 55303-7:1996 [88], which was used for many research projects and publications, is essentially equivalent.

and straightforward answer. The present section aims to help the designer in making an informed decision by presenting a choice of available information and data from research. For simplified approaches, standards, and regulations, the reader should refer to the design chapters (Chapters 4, 5 and 7).

The parameters that are used to describe the strength or the lifetime of structural glass elements vary among design methods. In the present section, the parameter set of the general lifetime prediction model described in Section 3.3 is used (*Table 3.3*). Its advantage over other commonly used parameter sets is that the parameters have a clear physical meaning and each include only one physical aspect.

The parameters related to crack growth and failure, namely the crack velocity parameters n and v_0, the crack growth threshold K_{th} and the fracture toughness K_{Ic} have already been discussed in Sections 3.2 and 3.3. The present section focuses on data concerning parameters that describe the flaws on glass surfaces, namely the surface condition parameters θ_0 and m_0 and the geometry factor Y.

Table 3.3: Overview of the parameters which influence the lifetime of structural glass elements.

Symbol	Designation	Main influence(s)
K_{Ic}	Fracture toughness	Material 'constant'
K_{th}	Crack growth threshold	Environmental conditions
Y	Geometry factor	Geometry of the crack and the element, stress field
v_0, n	Crack velocity parameters	Environmental conditions, stress rate
θ_0, m_0	Surface condition parameters (RSFP)	Glass surface condition
a_i	Initial depth of a surface crack (SSF)	Hazard scenario, glass type

3.6.2 Geometry factor

The geometry factor is in general a function of the stress field, the crack depth, the crack geometry, and the element geometry. This dependence can, however, generally be ignored for two reasons: Firstly, the crack growth that affects an element's lifetime occurs at crack depths that are very small in comparison with the element's thickness. Secondly, the depth and geometry of natural flaws are extremely variable. It does not make sense to increase the complexity of the model considerably to achieve a gain in accuracy that would be very small compared to these unavoidable uncertainties.

Table 3.4 gives some experimentally determined values for the geometry factor (non-italicized text). It remains, however, unknown to what extent these experiments represent actual flaws as they are encountered on glass elements under in-service conditions. Furthermore, it is difficult to separate the geometry factor's influence from other influences in tests, which is why experimental results must be interpreted with care.

To complement the experimental results, the geometry factor shall additionally be estimated using linear elastic fracture mechanics. Because of the extreme brittleness of glass, even elements that are exposed to very small loads fail immediately as soon as a surface crack grows to more than a few tenths of a millimeter. The following conditions of fracture mechanical relevance are, therefore, fulfilled in the case of macroscopic cracks on the surface of glass elements for structural engineering applications (for terminology,

Table 3.4: Overview of experimentally and analytically determined values for the geometry factor of surface flaws.

Type of flaw	Geometry factor Y
Glass on glass scratching [329, 330]	0.564
Vickers indentation [329, 330]	0.666
Half-penny shaped crack in a semi-infinite specimen	*0.637 – 0.713*†
Quarter-circle crack on glass edges [279]	0.722
Sandpaper scratching [329, 330]	0.999
Long, straight-fronted plane edge crack in a semi-infinite specimen	*1.120*

† Range of proposed values, see [189] for details.

see *Figure 3.4*): (a) The crack depth is small compared to the crack length. (b) The crack depth is small compared to the material thickness. (c) The radius of the crack front (not the crack tip) is substantially larger than the crack depth. (d) The crack depth is negligibly small compared to the overall dimensions of the structural element. This corresponds to the basic case of a *long, straight-fronted plane edge crack* in a semi-infinite specimen which has a geometry factor of $Y = 1.12$ [204]. This value was used by [44] and subsequently by all European work based on the DELR design method (cf. Chapter 4). Particularly flaws that are caused by hard contact are likely to be long. A geometry factor of $Y = 1.12$ seems, therefore, a sensible assumption for *surface cracks away from edges*.[23] This assumption is supported by the value of 0.999 that was determined for sandpaper scratching.

Surface flaws *on glass edges and at holes* depend mainly on the machining process and are likely to have different geometries from those that flaws on the surface have. There is, however, no quantitative data available that would shed some light on this issue. Based on theoretical considerations, [279] proposed modelling flaws on edges as quarter circle cracks with a geometry factor of $Y = 0.722$.

In order to put them into the context of the experimental values discussed before, the above-mentioned geometry factors are also listed in *Table 3.4* (italic text).

3.6.3 Ambient strength and surface condition

This text has been compiled in collaboration with the following experts:
Christoph HAAS

Glass surface away from edges

Existing glass strength data is difficult to compare. Firstly, past glass testing has been conducted at ambient conditions. The parameters that should represent the material strength depend, therefore, on the surface condition *and* on the crack growth behaviour during the tests. Secondly, the currently used 'strength parameters' are not material parameters but depend on the testing procedures used for their determination (cf. Section 3.5.2 and Chapter 4). It would not make much sense to compare the fracture strength of a small specimen which is exposed to an equibiaxial, linearly increasing stress (European tests) to

[23] Other researchers modelled glass surface damage as half-penny shaped cracks in a semi-infinite specimen ([170, 263, 287]). For such cracks, there is no single, universally used geometry factor. Popular solutions range from 0.637 to 0.713 [189].

the 60 s-equivalent uniformly distributed lateral load on a large rectangular glass plate (North American tests).

The second problem can be solved using the lifetime prediction model from Section 3.3. The surface condition parameters θ_0 and m_0 (cf. Equation (3.31)), which depend solely on the surface condition of the glass and are therefore true material parameters, can be derived from ambient strength data (for details and equations, see [189]). What cannot be avoided is the need to estimate the crack growth which takes place during ambient tests. The surface condition parameters shown in *Table 3.5* were calculated using $n = 16$, $v_0 = 0.01\,\text{mm/s}$, $Y = 1.12$, $K_{\text{Ic}} = 0.75\,\text{MPa}\,\text{m}^{0.5}$. The v_0 value is an estimation for laboratory tests at ambient conditions and medium stress rate, which was derived from existing data of experiments performed at various stress rates [189].[24]

For a discussion on how to use this data for design as well as on related problems, see Chapter 6.

Table 3.5: Surface condition parameters determined from laboratory tests at ambient conditions unless otherwise stated ($n = 16$, $v_0 = 0.01\,\text{mm/s}$).

	A (cm^2)	θ_0 (MPa)	m_0 (–)
As-received glass			
DIN 1249-10:1990 [80]	2400	62.89	4.94
Brown [48]		70.21	6.39
Beason and Morgan [33]		60.27	7.88
Fink [169]	23.8	61.20	6.30
Haldimann [189] from ORF data*		62.95	8.09
Haldimann [189] from inert tests*	20.4	67.57	7.19
Weathered window glass			
Beason [31]		27.65	5.25
ASTM E 1300-04 [22]		40.93	6.13
Fink [169]†	23.8	20.82	3.76
Glass with artificially induced homogeneous surface damage			
DELR/prEN 13474	2400	28.30	20.59
Blank [44]	2400	35.37	33.19
Blank [44]	2.54	33.29	23.53

* Inert conditions, therefore independent of the crack velocity parameters.
† The conversion of data from small specimens to the reference surface area of $A_0 = 1.0\,\text{m}^2$ tends to be unreliable for data with large scatter (small m_0).

[24] It is important to use a realistic and not a conservative value for v_0, because overestimating the crack growth during the tests means underestimating the surface damage on the specimen. The resulting surface flaw parameters would be too optimistic and thus unsafe.

The fact that the as-received glass parameters determined at ambient conditions are in close agreement with those determined from inert conditions suggests that the v_0 value used is a sensible choice.

Glass edges

Glass panels are cut to the required dimension with diamond cutters or water-jet cutters. For standard window glazing or insulated glass units, the resulting raw edge quality is sufficient. For heat-treated glass or structural glass elements, the glass edges have at least to be ground or polished and chamfered. Holes, which can be considered as curved glass edges, are drilled with a diamond borer or a water-jet cutter. The surface quality after drilling is comparable to a ground edge.

Due to machining, surface damage is generally more severe on glass edges than it is away from edges. The *inherent tensile edge strength* is, therefore, generally lower than the strength away from edges and depends strongly on the machining process and the edge finishing quality.[25]

Schneider [297] found that in annealed glass, holes drilled by water-jet have deeper surface flaws and therefore a lower tensile strength than holes drilled with diamond cutters. After tempering, he measured similar strengths for both drilling processes.

At present there is insufficient knowledge of the severity of the induced edge flaws such that it is difficult to predict the strength of glass elements on edges and at bolt holes accurately. Further investigations on the surface condition in function of the machining and finishing processes are required. On glass edges that are potentially exposed to severe damage during an element's lifetime, e. g. because of accidental impact or vandalism, this kind of damage is generally more critical than the machining damage and must, therefore, be considered for design (Chapter 6). Until more knowledge is available, the following is recommended: For protected glass elements, the published strength data or the values from ASTM E 1300 (see below) may be used. For exposed glass elements, a conservative assumption on the maximum flaw depth to be considered (design flaw) should be made. The strength of the design flaw can be estimated for instance with *Figure 3.6*.

At present, only edge strength data from tests at ambient conditions is available. It is, as mentioned before, difficult to compare since results depend on the testing procedure and the statistical procedures used to interpret the results. Many results are contradictory. The interested reader should refer to [41, 184, 237, 297]. ASTM E 1300 [22] is currently the only standard which specifies allowable tensile stresses for glass edges, see Section 4.4.2.

The influence of the surface condition and the residual stress on the compressive edge strength is small. It is, therefore, comparable to compressive strength away from edges. DIN 1249-10:1990 [80] gives values between 700 and 900 MPa, Wörner et al. [348] between 380 and 600 MPa. In experiments with various load introduction configurations, Luible [242] found compressive edge strength to be greater than 500 MPa.

[25] It should be noted that machining and edge finishing has not only an influence on the mean of the fracture strength, but also on its variability. The mean strength of polished glass edges, for instance, is higher than the one of edges with artificially induced homogeneous surface damage. However, when the characteristic strength is defined as the 5% fractile of some statistical distribution which is fitted to the results, it may well be lower for polished edges than for artificially damaged edges. What may seem paradox at first glance is just a consequence of the larger scatter of the polished edge data, which leads to a lower 5% fractile despite the higher mean value.

3.6.4 Residual surface stress due to thermal tempering

Away from edges

In contrast to the inherent glass strength, the residual surface stress does not depend on the surface condition, the loading history, or the environmental conditions. *Figure 3.9* shows published measurements, grouped by glass thickness.

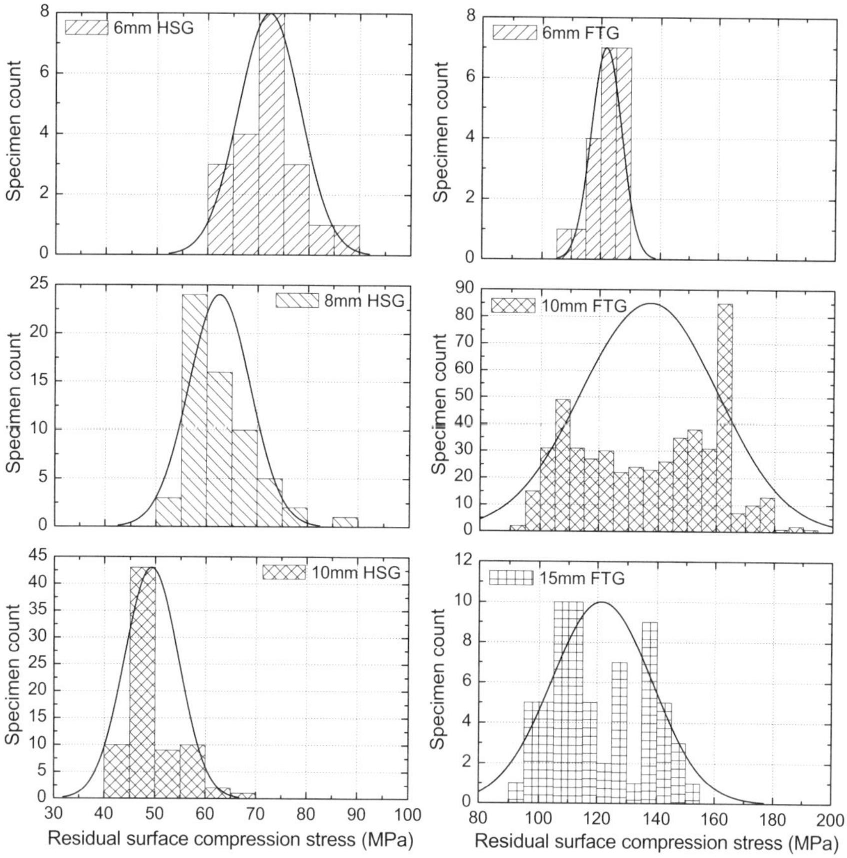

Figure 3.9: Residual stress data, pooled by glass thickness (histograms and fitted normal distributions; data sources: [189, 236, 243, 297]).

The residual stress data allows the following conclusions to be drawn:

- The residual surface stress varies widely among specimens as well as among manufacturers[26].
- The residual stress in heat strengthened glass seems to be inversely proportional to the glass's thickness.

[26] This is not directly visible in the figures, but is in the original data. It is also the reason why the fully tempered glass data look like a superposition of at least two samples with very different mean values.

- The characteristic residual stress for structural design can be defined as the 5%-fractile value of the normal distribution which fits to the pooled data of all thicknesses. With the data shown, the values obtained are $\sigma_{r,k,HSG} = 40\,\mathrm{MPa}$ for heat strengthened glass and $\sigma_{r,k,FTG} = 95\,\mathrm{MPa}$ for fully tempered glass. The values given in prEN 13474-1:1999 [280] are even lower, namely, 25 MPa for HSG and 75 MPa for FTG. This means that for the structural design of glass elements, a very large part of the residual stress is currently 'lost', i. e. cannot be considered for the design strength, because of the large scatter of the data. This 'lost strength' will often be a multiple of the long-term inherent glass strength (cf. *Figure 3.6*). Significantly more economical and aesthetic glass structures could, therefore, be designed if a high residual stress level could be guaranteed, e. g. by quality assurance measures.

Straight edges and holes

The residual stress distribution on edges and near holes is inhomogeneous and varies widely. It depends on the temperature distribution in the glass element during the tempering process, which is in turn a function of the element's geometry as well as of the cooling equipment and the cooling process.

Measurement of residual stresses on edges or near edges is difficult, time-consuming, and requires special equipment. It is therefore tempting to measure residual stresses away from edges and extrapolate these to obtain the stresses along edges. It should be noted though, that no distinct correlation between residual stress on edges and away from edges could be found [56]. In view of the above-mentioned dependance, this is not surprising.

Recent experimental and numerical investigations on glass strength and residual stress on edges and at holes were conducted, among others, by [41, 61, 237, 248, 297, 315]. Some of the main findings are summarized in the following:

- The tempering process can be simulated numerically, such that the residual stresses can be determined by simulation. Such simulations are, however, generally not practicable for design since they require advanced finite element software, expert knowledge about the physical processes and the complex simulations, and detailed information on the tempering facility.
- An alternative way of determining residual stresses around holes consists in experimentally measuring the strength at holes and subtracting the inherent glass strength. The procedure has, however, notable drawbacks. Firstly, the inherent strength at holes cannot be accurately determined. Secondly, the stress field due to loading is complex around holes. It needs to be calculated with FE models, which must often account for nonlinearities due to interfaces that transmit only compressive stresses. The results obtained from such FE models depend strongly on the model characteristics (mesh, element type, simplifying assumptions, material laws etc.).
- Residual stresses decrease towards the centreline of the glass pane. This effect is more pronounced with thick than with thin glasses.
- The residual stress on a glass surface reaches its minimum at about one to two glass thicknesses away from the edge. This effect, called 'overshoot', becomes more pronounced with increasing glass thickness. The analysis of failure origins in lateral torsional buckling tests showed that this phenomenon should be taken into account for structural design [243].

- In fully tempered glass, the residual stress on edges is significantly *lower* (15–25% on straight edges, 25–35% at holes) than it is away from the edges.
- In heat strengthened glass, however, the residual edge stress tends to be significantly *higher* (up to 50% on straight edges, about 15% at holes) than the residual stress away from the edges.
- The residual compressive stress near the chamfers of cylindrical holes is slightly higher than away from holes (approx. 10–15%), but residual stresses in holes decrease towards the centreline of the glass pane. This effect is more pronounced with thick glasses. With conical holes, the stress distribution is even more complex *Figure 3.10*.
- [237] and [297] obtained a number of results which are contradictory (although they are not directly comparable because different approaches were used for their determination). While residual stresses are about 10% smaller in holes than on straight edges according to Laufs, Schneider claims that they are higher in holes. For lateral loading and fully tempered glass, Schneider determined design strengths in holes which are even higher than those typically assumed away from edges, while Laufs determined much lower values. For in-plane loading, Laufs found strengths which are only about 50% of those typically assumed away from edges. Both researchers propose design values for the tensile strength in glass holes [237, 297].
- At a distance of about half the glass thickness away from holes, the residual stresses on the glass surface is slightly lower than it is away from edges [237, 297].
- Crack healing seems to have a strong beneficial effect on the edge strength of fully tempered glass. The effect is less pronounced with heat strengthened glass.

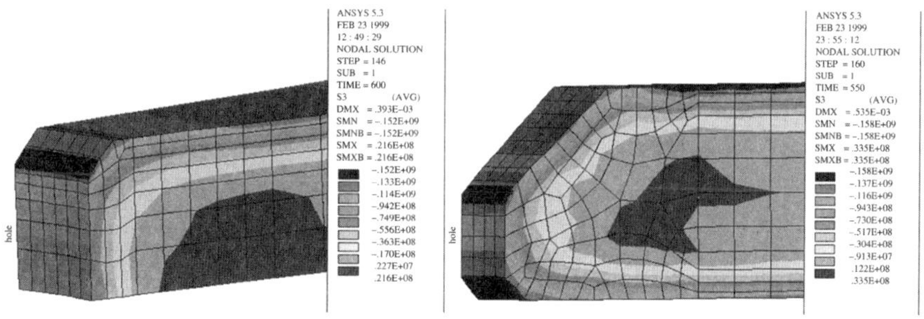

Figure 3.10: Simulated residual stress in a cylindrical (left) and a conical (right) hole [298].

Chapter

4 Current Standards, Guidelines and Design Methods

4.1 Introduction

The increasing use of glass as a load-bearing material has led to the development of a number of national and international design standards, draft standards, technical guidelines and recommendations. The aim of these documents is to arrive at an accurate value of allowable load or stress for an acceptable probability of failure in terms of the geometrical configuration of the glass (i. e. shape and support conditions) and the environmental parameters (loads and ambient conditions) by means of a few simple calculations.

These design methods do not cater for all types of glass configurations, loading, support and surface conditions. Most commonly, they are limited to glass elements of rectangular shape with continuous lateral support and to uniformly distributed out-of-plain loads. An in-depth analysis of the underlying assumptions in Section 4.5 reveals further limitations that the design methods fail to mention.

It is beyond the scope of this document to give an exhaustive overview of *all* national standards and design methods that exist in the field of glass. All the more, because many of them are based on simple theories, ignore geometrical non-linearity and the like. Although these methods are sufficiently accurate for rectangular window glazing with continuous lateral support, they should not be used for structural glass applications or for support and loading conditions that they do not cover. The standards and design methods discussed in the following have been chosen either because they are widely used or because they are of particular interest for structural glass design.

4.2 Rules of thumb

This text has been compiled in collaboration with the following experts:
Benjamin BEER

Accurate analysis and design methods are generally unattractive for manual computation and it is unrealistic to expect the engineer to perform laborious calculations throughout the whole of the iterative design process. This fuels the need for reliable rules of thumb for performing quick checks. Rules of thumb are a very useful tool for the structural engineer, but their use should be limited to scheme design purposes rather than as the basis for detailed design. Rules of thumb cannot replace detailed design. They simply help ensure that material selection, material quantity and consequently cost estimates are not too far from the final requirements. Furthermore, rules of thumb should be used as an approximate verification of the results obtained from detailed analysis.

4.2.1 Allowable stress based design methods

Despite the inaccuracy of this over-simplistic approach and the fact that the concept of allowable stress is rarely used in current building design standards, allowable stress design methods are still widely used to design glass elements. It is mainly the extreme ease of use and the simplicity of these methods that make them attractive. The general verification format is

$$\sigma_E \leq \sigma_{\mathrm{adm}} \tag{4.1}$$

σ_E maximum in-plane principal stress, calculated using the characteristic values of the actions of the most unfavourable design scenario;

σ_{adm} allowable principal in-plane stress (the fracture strength found in experiments, divided by a global safety factor that accounts for all uncertainties and variances associated with actions, resistance and modelling).

There is no way of considering the effects of the element's size, the environmental conditions, the duration of load and the like, or of taking a specific target failure probability into account. These aspects must all be somehow 'included' in the recommended σ_{adm} values.

The German technical guidelines TRLV 2006 [328] and TRAV 2003 [327] are well-known and widely used examples of design guides based on allowable stresses. Both the documents apply to only glass panes exposed to uniform lateral loads. The recommended allowable stresses for static loads are summarized in *Table 4.1*. For impact loads, TRAV 2003 [327] sets the following allowable stresses: 80 MPa for annealed glass (ANG), 120 MPa for heat strengthened glass (HSG) and 170 MPa for fully tempered glass (FTG).[1] Additionally, both the guidelines contain a series of more detailed specifications on how to account for load combinations. They also list special requirements that must be met and modified allowable stress values for a series of specific situations.

Allowable stresses have also been proposed for edges of glass beams, e. g. by Hess [194], Güsgen [184] and Laufs [237].

Table 4.2 shows allowable stresses recommended for initial design by Pilkington Glass Consultants [219, 277]. These values should only be used in conjunction with linear stress analysis.

Shortcomings

Allowable stress based design methods have some notable drawbacks:

[1] Note that these stresses are only valid for the impact load tests described in TRAV 2003 [327].

	Allowable stress σ_{adm} (MPa) Vertical glazing	Overhead glazing
Annealed glass (ANG)	18	12
Fully tempered glass (FTG)	50	50
Laminated ANG	22.5	15 (25*)

* Only for the lower glass pane in the hazard scenario 'upper pane broken'.

Table 4.1: Allowable stresses for glass panes exposed to uniform lateral load according to [328] and [327].

Table 4.2: Allowable stresses for initial design – recommendations by Pilkington.

Load type	Loading example	Annealed glass (MPa)	Fully tempered glass (MPa)
Short-term body stress	Wind	28*	59
Short-term edge stress	Wind	17.8*	59
Medium term	Snow	10.75	22.7
Medium term	Floors	8.4	35
Long term	Self weight, water, shelves	7	35

* Valid for annealed glass $\geq$ 10 mm. For 6 mm thick glass these values may be multiplied by a factor of 1.4.

- They do not account for the actual physical phenomena that govern the mechanical behaviour of glass.
- Scatter and uncertainty of the influencing parameters differ. With only one global safety factor, this cannot be accounted for.
- The approach is of very limited accuracy and flexibility and is not well suited to deal with aspects such as geometric non-linearity or instability.

4.2.2 Recommended span/thickness ratios

Table 4.3 provides a list of maximum unsupported spans proposed by Colvin [69] for initial design purposes of glazing with continuous lateral support along two or four edges. Furthermore, the maximum height of a vertical glass fin should not exceed 15× the fin depth [69].

Glass type	Maximum span/thickness ratio vertical	sloping or horiz.
Annealed glass (ANG)	150	100
Fully tempered glass (FTG)	200	150
Laminated ANG	150	100
Laminated FTG	150	100

Table 4.3: Maximum unsupported spans proposed by Colvin [69].

4.3 European standards and design methods

The rules of thumb described in Section 4.2 provide useful initial checks. It is, however, often necessary to carry out more accurate calculations in glass design. This section lists the more accurate design methods that are prevalent in Europe. These methods rely on an accurate determination of the tensile stresses in the glass, which in the presence of large bending deformations should involve a geometrically non-linear analysis (see Section 2.3.1).

4.3.1 DELR design method

Verification format

The design method of damage equivalent load and resistance, called DELR design method hereafter, was the first European glass design method that attempted to account for the specific behaviour of glass in an adequate and transparent way. It is compatible with the current generation of standards based on partial safety factors. Presented to a larger public in [302], the design method is based on research work by Richter [289], Kerkhof [226], Kerkhof et al. [227], Exner [166, 167], Blank [44], Güsgen [184] and others. Originally developed for glass plates, it was also extended to cover glass beams. The maximum principal design stress $\sigma_{\text{max,d}}$ is compared to an equivalent resistance as follows:

$$\sigma_{\text{max,d}} \leq \frac{\sigma_{\text{bB},A_{\text{test}},\text{k}}}{\alpha_{\sigma}(q,\sigma_{\text{V}})\cdot\alpha(A_{\text{red}})\cdot\alpha(t)\cdot\alpha(S_{v})\cdot\gamma_{\text{M,E}}} + \frac{\sigma_{\text{V,k}}}{\gamma_{\text{M,V}}} \tag{4.2}$$

$\alpha_{\sigma}(q,\sigma_{\text{V}})$ coefficient to account for the stress distribution on the glass surface; q = uniform lateral load[2], σ_{V} = residual surface stress due to tempering;

$\alpha(A_{\text{red}})$ coefficient to account for the size of the decompressed surface area[3] A_{red} (for annealed glass, A_{red} is equal to the entire surface area);

$\alpha(t)$ coefficient to account for the load duration;

$\alpha(S_{v})$ coefficient to account for load combination and environmental conditions;

$\sigma_{\text{max,d}}$ design value of the maximum in-plane principal stress in the element, calculated according to current action standards[4];

$\sigma_{\text{bB},A_{\text{test}},\text{k}}$ characteristic value of the inherent bending fracture strength in R400 coaxial double ring tests according to EN 1288-2:2000 [114] (see Section 3.5.2; 5% fractile, confidence level 0.95, surface area[5] $A_{\text{test}} = 0.24\,\text{m}^2$, stress rate $= 2 \pm 0.4\,\text{MPa/s}$);

$\sigma_{\text{V,k}}$ characteristic value (5% fractile) of the absolute value (compression = positive) of the residual surface stress (normally induced by thermal or chemical tempering called 'prestress' in the DELR design method);

[2] [302] uses p. q is used here for compatibility with the rest of the document.

[3] see Appendix B

[4] In particular, EN 1990:2002 [135], EN 1991-1-1:2002 [136], EN 1991-2-3:1996 [138], EN 1991-2-4:1995 [139], EN 1991-2-5:1997 [140], EN 1991-2-7:1998 [141] in connection with the National Application Documents in Europe; SIA 260:2003 [313] and SIA 261:2003 [314] in Switzerland.

[5] [302] uses A_0. This symbol is avoided here because it has a different meaning in the present document (unit surface area).

$\gamma_{M,E}$ partial factor for the inherent strength;

$\gamma_{M,V}$ partial factor for the residual stress.

Coefficients

A set of coefficients is used to compensate for the differences between laboratory test conditions (used to determine the strength) and actual in-service conditions. The *non-homogeneous stress distribution* on the glass surface is accounted for as follows:

$$\alpha_\sigma(q,\sigma_V) = \left[\frac{1}{A_{red}}\int_{A_{red}}\left(\frac{\sigma_1(x,y)}{\sigma_{max,d}}\right)^\beta \mathrm{d}x\mathrm{d}y\right]^{1/\beta} \tag{4.3}$$

$\sigma_1(x,y)$ is the major principal stress at the point (x,y) on the surface and depends (like A_{red}) on σ_V. The Weibull shape parameter, β, is assumed to be 25. This value has been defined by Blank [44] based on experiments on float glass samples with artificially induced homogeneous surface damage (sandblasting). It does not directly reflect the test data, but reflects a so-called 'limiting distribution' that lies somewhere below the test data. For standard cases, tabulated values of α_σ (from finite element calculations) are given, a simple but conservative assumption is $\alpha_\sigma = 1.0$.

The *size effect* is accounted for as follows:

$$\alpha(A_{red}) = \left(A_{red}/A_0\right)^{1/\beta} \tag{4.4}$$

The coefficient $\alpha(t)$ accounts for the *load duration*. It depends on the sub-critical crack growth, the duration of all loads in a load combination, the overlapping probability of wind load and snow load, the bending strength determined in tests, the stress rate used in these tests, the surface area and the required lifetime. For 'usual conditions' and a design life of 50 years, Sedlacek et al. [302] proposes to use $\alpha(t) = 3.9$.

The coefficient $\alpha(S_v)$ takes the *relative magnitude of the different loads* within a load combination as well as *environmental conditions* into account. Its calculation is complex and too lengthy to be discussed here; the interested readers are referred to [302]. The difference with respect to other design methods is that two sets of crack velocity parameters are used in the calculation of $\alpha(S_v)$: one for 'winter conditions' ($S_{Winter} = 0.82\,\mathrm{m/s(MPa\,m^{0.5})^{-n}}$) and one for 'summer conditions' ($S_{Summer} = 0.45\,\mathrm{m/s(MPa\,m^{0.5})^{-n}}$). $n = 16$ is assumed for both conditions.

Partial factors

In [302], a partial resistance factor of $\gamma_M \approx 1.80$ is proposed for structures of medium importance. This factor is chosen by a rather particular approach involving two Weibull distributions: First, a 'characteristic value of the inherent bending strength' $\sigma_{bB,A_{test},k} = 45\,\mathrm{MPa}$ is defined as the 5% fractile of a Weibull distribution with the parameters $\theta_{A_{test}} = 74$ MPa and $\beta_{test} = 6$. This distribution represents the breakage stress of as-received float glass specimens in an R400 coaxial double ring test (see Section 3.5.2) at a stress rate of $2 \pm 0.4\,\mathrm{MPa/s}$ and with $A_{test} = 0.24\,\mathrm{m^2}$ (95% confidence level). These tests were performed as a basis for DIN 1249-10:1990 [80]. Secondly a 'design bending strength of damaged specimens' $\sigma_{bB,A_{test},d} = 24.7\,\mathrm{MPa}$ is defined as the 0.0012 fractile value of the failure strength distribution proposed by Blank [44]. This distribution, characterized by $\theta_{A_{\text{test}},\text{limit}} = 32\,\text{MPa}$ and $\beta_{\text{limit}} = 25$, was chosen based on laboratory tests with the same setup as described above, but on specimens with artificially induced homogeneous surface damage. The chosen distribution is somewhat more conservative than the actual test results. The partial resistance factor is defined as $\gamma_{\text{M}} = \sigma_{\text{bB},A_{\text{test}},\text{k}}/\sigma_{\text{bB},A_{\text{test}},\text{d}} \approx 1.80$.

This approach is reused in unaltered form for the European draft code prEN 13474-1:1999 [280], see Section 4.3.2.

Extension for beams

Equation (4.2) is adapted for beams by adapting the coefficients:

$$\sigma_{\text{max,d}} \leq \frac{\sigma_{\text{bB,L}_{\text{test}},\text{k}}}{\alpha_\sigma(q,\sigma_{\text{V}})_{\text{BZ}} \cdot \alpha(L_{\text{red}}) \cdot \alpha_{\text{BZ}}(t) \cdot \alpha_{\text{BZ}}(S_v) \cdot \gamma_{\text{M,E}}} + \frac{\sigma_{\text{V,k}}}{\gamma_{\text{M,V}}} \tag{4.5}$$

$$\alpha_\sigma(q,\sigma_V)_{BZ} = \left[\frac{1}{L_{\text{red}}}\int_{L_{\text{red}}} \left(\frac{\sigma_1(l)}{\sigma_{\text{max,d}}}\right)^{\beta} \text{d}l\right]^{1/\beta} \qquad \alpha(L_{\text{red}}) = \left(\frac{L_{\text{red}}}{L_{\text{test}}}\right)^{1/\beta} \qquad \alpha(t) \approx 3.7 \tag{4.6}$$

$\sigma_{\text{bB},L_{\text{test}},\text{k}}$ is the characteristic bending strength (5% fractile) of beam specimens with decompressed length[6] L_{test} $(= 0.46\,\text{m})$. $\sigma_1(l)$ is the major principal stress at location l. $\alpha(L_{\text{red}})$ accounts for the length of a beam's decompressed edge. [302] recommends the use of $\beta = 5$ for polished and $\beta = 12.5$ for unpolished edges. The values were determined from very small samples (11 and 13 specimens, respectively). $\gamma_{\text{M,E}} \approx 1.40$ is proposed for $\beta = 12.5$. $\alpha_\sigma(q,\sigma_{\text{V}})_{\text{BZ}}$ equals 1.0 for a uniform stress distribution, 0.94 for a parabolic and 0.86 for a triangular one. $\alpha_{\text{BZ}}(S_v)$ is equal to $\alpha(S_v)$.

Shortcomings

There are no shortcomings that are very specific to this method. The more general ones are discussed in Section 4.5.

4.3.2 European draft standard prEN 13474

The design method of prEN 13474[7] [280, 281] is based on the DELR design method, but contains influences from the methods of Shen and Siebert (see Sections 4.3.3 and 4.3.4). The draft standard faced stiff opposition and is still under revision at the time of writing. The influence of the stress distribution on the glass surface is accounted for on the action side of the verification equation, the residual surface stress on the resistance side. The structural safety verification format compares an *effective stress*, σ_{eff}, with an *allowable effective stress for design*, $f_{\text{g,d}}$:

$$\sigma_{\text{eff,d}} \leq f_{\text{g,d}} \tag{4.7}$$

[6] The decompressed length is the length of the edge, where the tensile stress due to loading is greater than the residual compressive stress due to tempering.

[7] *Important*: This standard is currently under revision by the committees CEN/TC 250 ('Structural Eurocodes') and CEN/TC 129 ('Glass in Buildings'). At the time of writing, the non-public working papers differ considerably from the published draft standards [280] and [281].

The effective stress $\sigma_{\text{eff,d}}$[8] has to be determined for the most unfavourable action combination as follows:

$$\sigma_{\text{eff,d}} = \left[\frac{1}{A} \int_A \left(\sigma_1(x,y)\right)^\beta \, \mathrm{d}x\mathrm{d}y \right]^{1/\beta} \tag{4.8}$$

A is the *total* surface area of the glass pane and $\sigma_1(x,y)$ is the major principal stress *due to actions* at the point (x,y) on the surface. This means that the effective stress is defined *independently* from the residual stress and that decompression of the whole surface is assumed. Using the coefficient for annealed glass $\alpha_\sigma(p)$ from the DELR design method, it is $\sigma_{\text{eff,d}} = \sigma_{\text{max,d}} \cdot \alpha_\sigma(q)$. β is the shape parameter of the Weibull distribution of the breakage stress. For common geometries and support conditions, [281] provides tables and equations to determine $\sigma_{\text{eff,d}}$ in function of the applied load q and the plate dimensions without actually having to solve Equation (4.8).

The *allowable effective stress* is defined as follows:

$$f_{\text{g,d}} = \left(k_{\text{mod}} \frac{f_{\text{g,k}}}{\gamma_{\text{M}} \cdot k_A} + \frac{f_{\text{b,k}} - f_{\text{g,k}}}{\gamma_{\text{V}}} \right) \cdot \gamma_{\text{n}} \tag{4.9}$$

$f_{\text{b,k}}$ characteristic value of the fracture strength (5% fractile); $f_{\text{b,k}} = f_{\text{g,k}}$ for ANG, 70 MPa for HSG and 120 MPa for FTG

$f_{\text{g,k}}$ characteristic value of the inherent strength (5% fractile); $f_{\text{g,k}} = 45\,\text{MPa}$ for soda lime silica and borosilicate glass

$f_{\text{b,k}} - f_{\text{g,k}}$ contribution of residual stress to the failure strength (0 for annealed glass)

γ_{V} partial factor for the residual stress due to tempering (= 2.3 for SLS glass);

γ_{M} partial factor for the inherent strength (= 1.8 for SLS glass);

γ_{n} national partial factor (= 1.0 for most countries);

k_A coefficient to account for the surface area, defined independently from the residual stress as $k_A = A^{0.04}$ (from Equation (4.4) with $A_{\text{test}} = 1\,\text{m}^2$ and $\beta = 25$);

k_{mod} modification factor to account for load duration, load combination and environmental conditions; k_{mod} is given for the following dominant actions: short duration (wind), 0.72; medium duration (snow, climate loads for IGUs), 0.36; permanent loads (self weight, altitude for IGUs), 0.27.

In comparison to the DELR design method, prEN 13474 contains the following modifications:

- The factor to account for the influence of the stress distribution on the surface is defined independently from the residual stress.
- k_{mod} replaces $\alpha(t)$ and $\alpha(S_v)$.
- k_{A} replaces $\alpha(A_{\text{red}})$, but is based on the total instead of the decompressed surface area, which makes it independent of the residual stress. Additionally, it is defined with respect to a reference surface of $1\,\text{m}^2$ instead of $0.24\,\text{m}^2$. Surprisingly, it is used together with the unaltered characteristic strength value, which is based on $A_0 = 0.24\,\text{m}^2$.

[8] prEN 13474 does not use the index d, even when referring to the design level. The index is added here for clarity.

As a result of these modifications, the partial factors are not directly comparable. The replacement of $\alpha(t)$ and $\alpha(S_v)$ by k_{mod} is very similar to Shen's concept, but k_{mod} is not identical to η_D. Instead of explicitly accounting for the relative magnitudes of the different loads of a load combination, the few tabulated k_{mod} values include implicit assumptions. Appendix E of prEN 13474-2:2000 [281] proposes a step-by-step procedure for design, using predefined load combinations.

Shortcomings

The characteristic value of the inherent strength of float glass is said to be $f_{g,d} = 45\,\text{MPa}$. This value was originally defined in DIN 1249-10:1990 [80], based on coaxial double ring tests on new annealed glass specimens with a surface area of $A_{test} = 0.24\,\text{m}^2$ (cf. Section 3.5.2). A two-parameter Weibull distribution was fitted to the measured failure stresses. The Weibull parameters obtained were $\theta_{A_{test}} = 74\,\text{MPa}$ and $\beta = 6$ (at 0.95 confidence level). The characteristic value is defined as the 5% fractile value of this distribution, which gives the 45 MPa mentioned above. To account for the size effect, $f_{g,d}$ is divided by a size factor k_A defined as

$$k_A = A^{0.04} \tag{4.10}$$

with A being the total surface area of the glass plate. As discussed in Section 4.5, the actual size factor based on Weibull statistics is

$$k_{A,\text{Wb}} = \left(A/A_{test}\right)^{1/\beta} \tag{4.11}$$

A and A_{test} are the decompressed surface areas of the element to be designed and the specimen used to determine the characteristic strength. The following inconsistencies may therefore be identified:

- A characteristic resistance determined from a distribution with $\beta = 6$ is combined with a correction factor based on $\beta = 25$ (exponent $0.04 = 1/25$).
- The size factor k_A becomes 1 for $A = 1\,\text{m}^2$. This means that the surface area in the tests leading to $f_{g,d}$ is assumed to be approximately four times bigger than it actually was. For $\beta = 25$, the quantitative effect of this is relatively small. Using the real test surface $A_{test} = 0.24\,\text{m}^2$, it is $k_{A,\text{Wb}}(A = 1\,\text{m}^2) = 1.059$ (difference of 'only' 6%). For $\beta = 6$, however, it is $k_{A,\text{Wb}}(A = 1\,\text{m}^2) = 1.269$ (difference of 27%).

The more general shortcomings are discussed in Section 4.5.

4.3.3 Shen's design method

Shen presented this design method in [311]. In [348], it was adapted to the format of EN 1990:2002 [135]. It is mainly a considerable simplification of the DELR design method, with one important exception: the concept to account for residual stresses is taken from the Canadian Standard CAN/CGSB 12.20-M89 [60] (cf. Section 4.4.3). Only two coefficients, both on the resistance side of the verification equation, are used and very simple tables are proposed for their values. The residual surface stress of tempered glass is accounted for indirectly by these coefficients. The design method is confined to laterally loaded glass panes, made of annealed or fully tempered glass, with continuous lateral support along all four edges. An application to structural elements such as beams or columns is not immediately possible. The structural safety verification format is as follows:

$$\sigma_{\text{max,d}} \leq \sigma_{\text{k}} \cdot \frac{\eta_{\text{F}} \cdot \eta_{\text{D}}}{\gamma_{\text{R}}} \tag{4.12}$$

$\sigma_{\text{max,d}}$ design value of the maximum principal stress;

σ_{k} characteristic value of the bending strength determined in R400 coaxial double ring tests (cf. Section 3.5.2);

η_{F} coefficient to account for surface area stress distribution;

η_{D} coefficient to account for load duration;

γ_{R} partial factor for the resistance.

The verification has to be done separately for every different load duration. The factor η_{F} for the surface area and the stress distribution is defined in a simplistic way, see *Table 4.4*. The load duration factor η_{D} is a function of the glass type and is given in *Table 4.5*. To derive these values, the surface condition and the environmental conditions in structural applications have been assumed to be identical to those in the bending strength laboratory tests. With this assumption, it is

$$\eta_{\text{D}} = \frac{\sigma_{\text{D}}}{\sigma_{\text{R}}} = \left[\frac{t_{\text{R}}}{t_{\text{D}}} \cdot \frac{1}{n+1}\right]^{\frac{1}{n}} \tag{4.13}$$

σ_{D} equivalent strength;

σ_{R} bending strength found in laboratory tests (cf. Footnote 9);

t_{R} test duration[9];

t_{D} load duration;

n crack velocity parameter; $n_{\text{ANG}} \approx 17$ (annealed glass), $n_{\text{FTG}} = 70$ (fully tempered glass).

	$A = 0.5$–$4.0\,\text{m}^2$	$A = 4$–$10\,\text{m}^2$
ANG	1.0	0.9
FTG	1.0	1.0

Table 4.4:
Factor η_F for Shen's design method [348].

	Dead load (50 years)	Snow (30 days)	Wind (10 min)
ANG	0.27	0.45	0.69
FTG	0.74	0.83	1.00

Table 4.5:
Factor η_D for Shen's design method [311].

[9]The bending strength values from DIN 1249-10:1990 [80], which refer to tests with a stress rate of 2 MPa/s, are used: $t_{\text{R,ANG}} = 45\,\text{MPa}/2\,\text{MPa/s} = 22.5\,\text{s}$, $t_{\text{R,FTG}} = 120\,\text{MPa}/2\,\text{MPa/s} = 60\,\text{s}$

Shortcomings

The value $n_{\mathrm{FTG}} = 70$ is taken from CAN/CGSB 12.20-M89 [60] without further discussion. It has two drawbacks:

1. Increasing a crack velocity parameter is unrelated to the actual physical phenomena governing the resistance of heat-treated glass (cf. Section 3.3.4).
2. Although the value of 70 is indeed given in CAN/CGSB 12.20-M89 [60], it does not relate to the crack velocity parameter n in this standard, despite the use of the symbol n. It is actually the value of a parameter combining n with a constant for the relationship between lateral load and stress in rectangular plates (see Section 4.4.3).

For a combination of loads of different duration, η_{D} has to be calculated individually. [311] makes proposals on how this should be done for the combination of snow and dead load as well as for snow and wind.

The choice of the partial factor γ_{R} for the resistance depends on the target reliability level and the scatter of the bending strength data. Based on the assumption that the bending strength's coefficient of variation is 0.1, [311] proposes $\gamma_{\mathrm{R}} \approx 1.25$ for buildings of medium importance.[10] Wörner et al. [348] provide no value for γ_{R}.

The more general shortcomings are discussed in Section 4.5.

4.3.4 Siebert's design method

This design method was proposed by *Siebert* in [316]. The major modifications with respect to the aforementioned methods are as follows:

- An approach to account for the influence of biaxial stress fields is proposed.
- The residual stress is considered as an action.

The structural safety verification format is

$$\sigma_{\mathrm{ges,d,max}} \cdot f_{\mathrm{A}} \cdot f_{\sigma} \cdot f_{\mathrm{tS}} \leq \frac{\theta}{f_{\mathrm{P}}} \tag{4.14}$$

$\sigma_{\mathrm{ges,d,max}}$ maximum principal surface stress ($\sigma_{\mathrm{ges,d,max}} = \sigma_{\mathrm{d,max}} + \sigma_E$);

$\sigma_{\mathrm{d,max}}$ maximum principal stress due to actions;

σ_E residual surface stress (compression ⇒ negative sign);

f_{A} coefficient to account for the different surface areas of test specimen and actual structural element;

f_{σ} coefficient to account for the different stress distributions in the test specimen and the actual structural element;

f_{tS} coefficient to account for load duration and relative magnitudes of different loads;

θ scale parameter of the Weibull distribution fitted to experimental bending strength data (has the dimension of a stress);

f_{P} factor to account for the target failure probability.

[10] To find this γ_{R}, the resistance is assumed to follow a log-normal distribution. This assumption is incompatible with the size effect, which is a direct consequence of Weibull statistics (cf. Chapter 3). [311] does not comment on this issue.

The *stress due to action* is calculated as in the DELR design method. The residual stress $\gamma_V \sigma_V$, however, is considered as an action. Its partial factor cannot be defined on a firm scientific basis due to the lack of data. Siebert proposes $\gamma_V = 1.25$, which conceptually means putting the residual stress back to the resistance side. For a favourable action, the partial factor should rather be < 1.0, which is $1/\gamma_V$.

To obtain *resistance data*, Siebert recommends standard R400 coaxial double ring tests according to EN 1288-2:2000 [114] on specimens with artificially induced homogeneous surface damage and data analysis according to DIN 55303-7:1996 [88], see Section 3.5.2. If tests are performed on heat-treated glass, the residual stress has to be deduced from the breaking stress. Siebert proposes, however, to use annealed glass for testing because (a) measurement of the residual stress is imprecise and (b) defects caused by a given method of artificial damaging are more severe in annealed than in heat-strengthened or fully tempered glass.

To account for a non-homogeneous stress distribution within the element, the use of a so-called *effective area* $A_{N,ef}$ is proposed:

$$A_{N,ef} = \int_A \left(\frac{\chi \cdot \sigma_{ges,d}(x,y)}{\sigma_{ges,d,max}} \right)^{\beta} dA \tag{4.15}$$

$\sigma_{ges,d}(x,y)$ first principal design stress at the point (x,y) on the surface (this refers to the crack opening stress $\Rightarrow \sigma_{ges,d}(x,y) \geq 0$);

$\sigma_{ges,d,max}$ maximum first principal design stress on the surface;

A surface area of the glass pane;

χ correction factor for the ratio of major and minor principal stress; (conservative assumption: 1.0; for a uniaxial stress field, $\chi \approx 0.83$ is proposed).

Using $A_{N,ef}$, a coefficient to account for the difference in surface areas of test specimens and actual structural elements is defined as

$$f_{A\sigma} = \left(\frac{A_{N,ef}}{A_{L,ef}} \right)^{1/\beta} \tag{4.16}$$

$A_{L,ef}$ is the effective area of the test specimen. To simplify design tables, it is proposed to split $f_{A\sigma}$ into two factors as follows:

$$f_A = \left(\frac{A}{A_{L,ef}} \right)^{1/\beta} \qquad f_\sigma = \left(\frac{A_{N,ef}}{A} \right)^{1/\beta} = \frac{\sigma_{ges,d,ef}}{\sigma_{ges,d,max}} \tag{4.17}$$

The effective principal stress $\sigma_{ges,d,ef}$ is defined such that $A \cdot \sigma_{ges,d,ef}^{\beta} = A_{eff} \cdot \sigma_{ges,d,max}^{\beta}$. As residual stress is considered as an action, f_σ depends on it. f_A is identical to $\alpha(A)$ in the DELR design method.

The load duration, the relative magnitude of different loads in a load combination and the environmental conditions are accounted for by the factor f_{tS}, which is the product of the factors $\alpha(t)$ and $\alpha(S_v)$ from [184] ([316] uses identical assumptions and equations). An additional factor, f_P, enables a target probability of failure G_a to be chosen. It is defined as

$$f_{\mathrm{P}} = \left[\ln\left(\frac{1}{1-G_{\mathrm{a}}}\right)\right]^{-1/\beta} \tag{4.18}$$

[316] uses the failure probabilities proposed by [44] and [184]. According to these, it is $1.5 \cdot 10^{-3}$ for structures of medium importance, which gives $f_{\mathrm{P}} = 1.30$ when using $\beta = 25$ from [44].

In comparison with that used in the DELR design method, Siebert's partial factor for the residual compressive surface stress is clearly less conservative. For annealed glass, both methods give basically identical results despite the different partial factors: $\sigma_{\mathrm{Rd}} = \sigma_{\mathrm{bB,A_{test},k}}/\gamma_{\mathrm{M}} = 45\,\mathrm{MPa}/1.8 \approx \theta/f_{\mathrm{P}} = 32\,\mathrm{MPa}/1.3 \approx 25\,\mathrm{MPa}$. The reason for the different factors is that the resistance is based on the Weibull scale parameter (θ) in Siebert's method and on the characteristic strength ($\sigma_{\mathrm{bB},A_{\mathrm{test}},\mathrm{k}}$) in the DELR design method.

Shortcomings

There are no shortcomings that are very specific to this method. The more general ones are discussed in Section 4.5.

4.4 North American standards and design methods

This section lists the more accurate design methods that are prevalent in North America. In a similar fashion to the European methods discussed in Section 4.3, these methods rely on an accurate determination of the tensile stresses in the glass, which in the presence of large bending deformations should involve a geometrically non-linear analysis (see Section 2.3.1).

4.4.1 Glass failure prediction model (GFPM)

The glass failure prediction model (GFPM) presented in [31] and [33] is directly based on the statistical theory of failure for brittle materials advanced by [336]. According to Weibull, the failure probability of a brittle material can be represented as

$$P_{\mathrm{f}} = 1 - e^{-B} \tag{4.19}$$

where B reflects the risk of failure as a function of all relevant aspects, in particular the surface condition and the stress distribution. For general cases, the GFPM proposes the risk function

$$B = \tilde{k} \int_A \left[\tilde{c}(x,y)\sigma_{\mathrm{eq,\,max}}(q,x,y)\right]^{\tilde{m}} \,\mathrm{d}A \tag{4.20}$$

in which $\tilde{c}(x,y)$ is the 'biaxial stress correction factor' (a function of the minor to major principal stress ratio), A the surface area and $\sigma_{\mathrm{eq,\,max}}(q,x,y) = \sigma(q,x,y)(t_{\mathrm{d}}/60)^{1/16}$ the maximum equivalent principal stress as a function of the lateral load q and the point on the plate surface (x,y). $\tilde{m}$ and $\tilde{k}$ are the so-called 'surface flaw parameters'.[11] Based

[11] The tildes are not used in the source. They are required in the present document to avoid confusion in subsequent chapters.

on this, the following expression is introduced for rectangular glass plates exposed to uniform lateral loads of constant duration:

$$B = \tilde{k}(ab)^{1-\tilde{m}} \left(Eh^2\right)^{\tilde{m}} \left(\frac{t_\mathrm{d}}{60}\right)^{\tilde{m}/16} \tilde{R}\left(\tilde{m}, \tilde{q}, \frac{a}{b}\right) \tag{4.21}$$

a and b are the rectangular dimensions of the plate ($a > b$), h is the effective thickness, t_d is the load duration in seconds and E is Young's modulus (71.7 GPa in the GFPM). The non-dimensional function

$$\tilde{R}\left(\tilde{m}, \tilde{q}, a/b\right) = \frac{1}{ab} \int_A \left[\tilde{c}(x,y)\tilde{\sigma}_{\max}(\tilde{q},x,y)\right]^{\tilde{m}} \,\mathrm{d}A \tag{4.22}$$

depends on the surface flaw parameter $\tilde{m}$ and the distribution of the non-dimensionalized stress on the surface. $\tilde{q} = q(ab)^2/(Eh^4)$ is the non-dimensionalized load and $\tilde{\sigma} = \sigma(q,x,y)ab/(Eh^2)$ is the non-dimensionalized stress.

Shortcomings

The surface flaw parameters $\tilde{m}$ and $\tilde{k}$ cannot be measured directly. They are determined from constant load rate tests on rectangular glass plates using a rather complex iterative procedure. In order to establish the stress/time relationship at the location of the critical flaw (i. e. the flaw that caused failure), the failure origin has to be determined visually. From this relationship, the 60 seconds equivalent failure stress and the corresponding 60 seconds equivalent failure load is calculated. Then, a set of risk factors, $\tilde{R}\left(\tilde{m}, \tilde{q}, a/b\right)$, corresponding to each equivalent failure load is calculated for a wide range of assumed $\tilde{m}$. The best value of $\tilde{m}$ is determined by choosing the one which results in a coefficient of variation of the risk factor closest to 1.0. $\tilde{k}$ can then be calculated using the plate's geometry and the mean of the set of $\tilde{R}\left(\tilde{m}, \tilde{q}, a/b\right)$ for the best $\tilde{m}$. Both its magnitude and units are dependent on $\tilde{m}$.

Some minor improvements of the GFPM and its implementation in ASTM E 1300 are presented in [32] and integrated into recent versions of the standard.

The more general shortcomings are discussed in Section 4.5.

4.4.2 American National standard ASTM E 1300

The American National standards 'Standard Practice for Determining Load Resistance of Glass in Buildings' ASTM E 1300-04 [22] provides extensive charts to determine the required thickness of glass plates. It is based on the glass failure prediction model by Beason and Morgan (see Section 4.4.1) and on the finite difference stress and deflection analysis by Vallabhan [331]. Resistance is defined using a target failure probability of 0.008. ASTM E 1300 applies to vertical and sloped glazing in buildings exposed to a uniform lateral load and made of monolithic, laminated or insulating glass elements of rectangular shape with continuous lateral support along one, two, three or four edges. The specified design loads may consist of wind load, snow load and self-weight with a total combined magnitude less than or equal to 10 kPa. The standard does *not* apply to other applications such as balustrades, glass floor panels and structural glass members or to any form of wired, patterned, etched, sandblasted, drilled, notched or grooved glass or to any glass with surface and edge treatments that alter the glass strength. The verification format is

$$q \leq \mathrm{LR} = \mathrm{NFL} \cdot \mathrm{GTF} \tag{4.23}$$

with q being the uniform lateral load, LR the 'load resistance', NFL the 'non-factored load' (based on a 3 seconds load duration) and GTF the so-called 'glass type factor' (load-duration dependent, see below).

The important difference with respect to European design methods is that this verification format is based on *loads* and not on *stresses*. Furthermore, it does not use any partial factors. The NFL is determined from charts given for various geometries, support conditions, glass thicknesses and for monolithic as well as laminated glass. The GTF combines glass type and load duration effects and is given for single panes (*Table 4.6*) as well as for insulating glass units.

Table 4.6: Glass type factors (GTF) for a single pane of monolithic or laminated glass.

Glass type	Short-duration load	Long-duration load
ANG	1.0	0.5
HSG	2.0	1.3
FTG	4.0	3.0

All charts and values are calculated using the GFPM with $\tilde{m} = 7$, $\tilde{k} = 2.86 \cdot 10^{-53}\,\mathrm{N}^{-7}\,\mathrm{m}^{12}$, a Young's modulus of $E = 71.7\,\mathrm{GPa}$ and the effective (not the nominal) glass thickness [22, 32]. The non-factored load charts incorporate the viscoelastic model for the plastic interlayer from [39]. This model claims to describe accurately the evolution of the polymer shear modulus at 50 °C. At this temperature and for a load duration of 3 seconds (the reference in the standard), the PVB interlayer is characterized with an effective Young's modulus of 1.5 MPa. This value is meant to be a lower bound for commercially available PVB interlayers.

For independent stress analyses required in the case of special shapes or loads not covered in the standard, allowable surface stresses for a 3 seconds duration load are given, see *Table 4.7*. The values for edges are taken from [335]. It is claimed for the allowable 3 seconds stress and $P_\mathrm{f} < 0.05$ in annealed glass away from the edges that the following equation should give conservative values:

$$\sigma_{\mathrm{allowable}} = \left(\frac{P_\mathrm{f}}{\tilde{k}\,(d/3)^{7/n} A} \right)^{1/7} \tag{4.24}$$

The constant 7 in Equation (4.24) is the parameter $\tilde{m}$ and 3 is the reference time period in seconds. For $P_\mathrm{f} = 0.008$, $d = 3$ seconds and $A = 1\,\mathrm{m}^2$, Equation (4.24) yields 16.1 MPa, which is indeed very conservative with respect to the value of 23.3 MPa given in *Table 4.7*.

To be able to compare the allowable stresses in *Table 4.7* to those from *Table 4.1*, they must be converted to the same reference time period ($\sigma_{60\mathrm{s}} = \sigma_{3\mathrm{s}}(3/60)^{1/16} = \sigma_{3\mathrm{s}} \cdot 0.829$). For annealed glass, very similar values are obtained. For fully tempered glass, the allowable stress is clearly higher according to ASTM E 1300-04 [22] ($\sigma_{60\mathrm{s}} = 77.2\,\mathrm{MPa}$) than according to TRLV 2006 [328] ($\sigma_{60\mathrm{s}} = 50\,\mathrm{MPa}$).

	Annealed glass	**Heat-strengthened glass**	**Fully tempered glass**
Away from the edges	23.3	46.6	93.1
Clean cut edges	16.6	n/a	n/a
Seamed edges	18.3	36.5	73.0
Polished edges	20.0	36.5	73.0

Table 4.7: Allowable surface stress (MPa) for a 3 seconds duration load according to ASTM E 1300-04 [22].

The 3 seconds duration load that represents the combined effects of I loads of different duration (all normal to the glass surface) is determined using[12]

$$q_3 = \sum_{i=1}^{I} q_i \left[\frac{d_i}{3} \right]^{1/n} \tag{4.25}$$

where q_3 is the magnitude of the 3 seconds duration uniform load and q_i the magnitude of the load having duration d_i. For annealed glass, $n = 16$.

Shortcomings

Haldimann [189] showed that even if all explicit and implicit simplifying assumptions behind the design concept in ASTM E 1300 are considered, Equation (4.25) (Equation (X7.1) in ASTM E 1300-04 [22]) is not correct. It should read as follows (cf. Equation (3.57)):[13]

$$q_3 = \left(\frac{1}{3\,\mathrm{s}} \sum_{i=1}^{I} \left[q_i^n \cdot d_i \right] \right)^{1/n} \tag{4.26}$$

The more general shortcomings are discussed in Section 4.5.

4.4.3 Canadian National standard CAN/CGSB 12.20

The Canadian National standard 'Structural Design of Glass for Buildings' CAN/CGSB 12.20-M89 [60] deals with soda lime silica glass panes exposed to uniform lateral load. Like the American National standard, it is based on the GFPM (see Section 4.4.1) and a target failure probability of $P_f = 0.008$ for the resistance. It is important to notice that in contrast to ASTM E 1300-04 [22], which uses a 3 seconds reference duration for the resistance, CAN/CGSB 12.20-M89 [60] is based on 60 seconds reference duration. This is due to the fact that the Canadian standard, published in 1989, is based on ASTM E 1300-94 [23] while the 3 seconds reference duration was only introduced in ASTM E 1300-03 [21].

[12] Caution, this equation is incorrect. See the excursus below.

[13] This view is supported by a simple example in the following. There should obviously be no difference between loading a glass element with 10 kN once for 60 s and loading it with the same load twice for 30 s. When using Equation (4.26), the result is indeed the same for both cases: $q_{3s} = 12.06\,\mathrm{kN}$. Equation (4.25), however, yields $q_{3s} = 10\,\mathrm{kN} \cdot (60\,\mathrm{s}/3\,\mathrm{s})^{1/16} = 12.06\,\mathrm{kN}$ for the first case and $q_{3s} = 10\,\mathrm{kN} \cdot (30\,\mathrm{s}/3\,\mathrm{s})^{1/16} + 10\,\mathrm{kN} \cdot (30\,\mathrm{s}/3\,\mathrm{s})^{1/16} = 23.10\,\mathrm{kN}$ for the second case.

Standard cases

For standard cases, the verification format is as follows:

$$E_\mathrm{d} \leq R_\mathrm{d} \tag{4.27}$$

E_d combination of all actions (design level = including partial factors);

R_d resistance of the pane (design level = including partial factors).

The action term is

$$E_\mathrm{d} = \alpha_D D + \gamma \cdot \psi \cdot (\alpha_L L + \alpha_Q Q + \alpha_T T) \tag{4.28}$$

D dead loads (self weight, invariant hydrostatic pressure);

L live load (snow, rain, use and occupancy, variable hydrostatic pressure);

Q live loads (wind, stack effect, earthquake, climatic and altitude load for IGUs);

T effects of temperature differences except those included in Q;

α_x partial factors: $\alpha_D = 1.25$ (unfavourable) or 0.85 (favourable) $\alpha_L = \alpha_Q = 1.50$ $\alpha_T = 1.25$;

γ importance factor: $\gamma = 1.0$ (in general), $\gamma \geq 0.8$ (farm buildings having low human occupancy or buildings for which collapse is not likely to cause serious consequences);

ψ load combination factor: $\psi = 1.0$ (when only one of L, Q and T acts), $\psi = 0.7$ (when two of L, Q and T act), $\psi = 0.6$ (when all of L, Q and T act). The combination with the most unfavourable effect has to be determined.

The resistance term is

$$R = c_1 \cdot c_2 \cdot c_3 \cdot c_4 \cdot R_\mathrm{ref} \tag{4.29}$$

c_1 glass type factor: 1.0 (flat glass, laminated glass), 0.5 (sand blasted, etched or wired glass);

c_2 heat treatment factor: 1.0 (annealed glass), 2.0 (heat-strengthened glass), 4.0 (fully tempered glass);

c_3 load duration coefficient

load type	approx. equiv. duration	ANG	HSG	FTG
wind and earthquake	1 min	1.0	1.0	1.0
sustained (snow, ponding)	1 week to 1 month	0.5	0.7	0.9
continuous (dead load, hydr. pressure)	1 year to 10 years	0.4	0.6	0.8

c_4 load sharing coefficient (for insulating glass units): 1.0 (monolithic glass), 1.7 and 2.0 (double-glazed and triple-glazed sealed insulating glass units with similar glass types and thicknesses);

R_ref reference factored resistance of glass (the standard gives tabulated values) (factored resistance of annealed glass loaded to failure under a constant load in 60 s; the values given are based on the minimum allowable (not the nominal) thickness and an expected failure probability of 0.8%).

Laminated glass may be considered as monolithic glass if the load duration is $<$ 1 minute and the temperature $<$ 70 °C or if the load duration is $<$ 1 week and the temperature $<$ 20 °C. For any other condition, laminated glass has to be considered as layered glass (no composite action may be assumed).

Special cases

For non-standard applications that are not covered by the tables and factors, some more general indications are given. They allow to get more insight into the model that the tabulated values are based on. The area effect is accounted for by

$$R_A = R_{\mathrm{ref}} \cdot A^{(-1/\widetilde{m})} \tag{4.30}$$

where A is the area of the pane in m^2 and $\widetilde{m}$ 'varies from about 5 to 7'. The load duration effect is accounted for by

$$R_t = R_{\mathrm{ref}} \cdot t^{(-1/\tilde{n})} \tag{4.31}$$

with t being the load duration in minutes and $\tilde{n}$ being 15 for ANG, 30 HSG and 70 for FTG.

It is crucial to notice that $\tilde{n}$ is *not* equal to the exponential crack velocity parameter n although the letter 'n' is used in CAN/CGSB 12.20-M89 [60]. The tilde has therefore been added here to avoid confusion. Based on [221, 222], the standard assumes that $\sigma \propto R^c$, σ being the 'stress in fracture origin areas', R the uniform lateral load and c a constant < 1. It is $\tilde{n} = cn$, which means that Equation (4.31) is in fact a combination of Brown's integral (see Section 3.3.4) with the proportionality between the stress and q^c found for rectangular plates (q is the uniform lateral load). As this proportionality and the value of c arc included in $\tilde{n}$, the tables and equations in the Canadian standard should not be applied to other geometries, boundary conditions or loading conditions. The value of c is not directly given in the standard, but as n is said to be 16 (called d in the terminology of the standard) and $\tilde{n} = 15$, it should be $15/16$.

For general cases, CAN/CGSB 12.20-M89 [60] recommends to limit stresses to 25 MPa away from the edges of plates and to 20 MPa on clean-cut edges. These values have to be corrected by the factor for the area effect and most probably also for the load duration, although the latter is not mentioned explicitly.

Shortcomings

The n versus $\tilde{n}$ issue gives rise to a certain number of problems and misunderstandings. This has already been seen when discussing Shen's design method (Section 4.3.3), but it also affects the Canadian standard itself. In Appendix B of the standard, the 1-minute reference resistance R_{ref} is said to be (R_{f} is the failure load at the time of failure t_{f} in minutes):

$$R_{\mathrm{ref}} = R_{\mathrm{f}} \left(\frac{t_{\mathrm{f}}}{\tilde{n}+1} \right)^{1/\tilde{n}} \tag{4.32}$$

Using the equations in Section 3.3.4 and $\sigma \propto R^c$, it is seen, however, that it should be:[14]

$$R_{\mathrm{ref}} = R_{\mathrm{f}} \left(\frac{t_{\mathrm{f}}}{n+1} \right)^{1/\tilde{n}} \tag{4.33}$$

The more general shortcomings are discussed in Section 4.5.

[14] Though this has already been pointed out in [170], the standard has not yet been revised at the time of writing.

4.5 Analysis and comments

The preceding sections have shown that most of today's design methods are actually variations, extensions or simplifications of others. These may be grouped into *European design methods*, which are based on the DELR design method, and *North American design methods*, which are based on the GFPM.

The following analyses and comments aim at helping the engineer, who uses the above-mentioned design methods, to better understand their bases, advantages, drawbacks and limits of validity.

Main concepts

In *Table 4.8*, European and North American design methods are compared with respect to the main concepts that they are based on. It can be seen that the two approaches are not directly comparable because of conceptual incompatibilities. Data from experiments designed for one method cannot directly be used in conjunction with the other method.

Time-dependence of the glass strength

The tensile strength of glass is time dependent due to stress corrosion (see Section 3.2). The design methods presented above conceal this dependence within coefficients, such that the underlying assumptions are not readily visible. They are, therefore, briefly discussed in the ensuing text.

Current glass design methods assume that the crack velocity parameters are well-known constant values. The 'classic' European crack velocity parameters have been published in [227]. They are based on the ambient condition crack growth data from Richter [289], who determined the parameters by optically measuring the growth of large through-thickness cracks on the edge of specimens loaded in uniform tension. On this basis, 'design parameters' for the DELR design method were chosen in [44]. (These design parameters represent substantially higher crack velocities than Richter's measurements, see *Figure 3.3*.) The European draft standard prEN 13474 and the design method by Siebert are directly based on the DELR design method and use the same parameters. Shen uses a different approach, see Section 4.3.3.

The GFPM uses only one crack velocity parameter explicitly. It is equivalent to the parameter n in European design methods and assumed to be equal to 16.

Laboratory testing to obtain strength data

The design methods presented in this chapter are based on strength data obtained at ambient conditions. The parameters meant to represent the surface condition or a 'material strength' ($\tilde{k}$ and $\tilde{m}$ in the North American design methods, θ_A and β in the European design methods) are, therefore, inevitably dependent on the surface condition *and* on crack growth behaviour. This is a drawback for two reasons. Firstly, unrelated physical aspects are combined within a single value. Secondly, the large scatter and the stress rate dependence of the crack velocity parameters (see Section 3.2) make accurate estimation of the crack growth that occurs during experiments at ambient conditions difficult. Inaccurate estimation, however, can yield unsafe results. This issue is further discussed in Chapter 6.

Table 4.8: Comparison of European and North American design methods.

	European design methods	**North American design methods**
Testing procedure to obtain strength data	Constant stress rate coaxial double ring tests with $\dot{\sigma}_{\text{test}} = 2 \pm 0.4\,\text{MPa/s}$ on specimens with $A_{\text{test}} = 0.24\,\text{m}^2$ (cf. Section 3.5.2).	Large rectangular glass plates exposed to uniform lateral load.
Surface condition of strength test specimens	Artificially induced homogeneous surface damage.*	Weathered windows glass.
Design 'strength' definition	A single value is used, the 'characteristic value of the inherent strength'†. It is defined as the 5% fractile value (at 0.95 confidence level) of the failure stress measured in the experiments.	Two interdependent parameters called 'surface flaw parameters' $\tilde{m}$ and $\tilde{k}$ are used. Their determination from experimental data is based on the stress history at the visually determined failure origin and a rather complex iterative procedure.
Sub-critical crack growth	Taken into account by load duration factors that depend on the loading only. The factors are based on the empirical relationship $v = S \cdot K_I^n$ (cf. Section 3.2).	Only one crack velocity parameter is used explicitly. It is equivalent to the parameter n in European methods and assumed to be 16.
Extrapolation from experiments to in-service conditions	Some but not all differences between laboratory conditions and actual in-service conditions are accounted for by correction coefficients. Details vary between methods, see Section 4.3.	Graphs are provided for many common cases (in terms of geometry and support conditions). They provide uniform lateral loads that a given glass pane can withstand for a reference time period.
Taking the glass type into account	Mostly by adding the absolute value of the residual surface stress (multiplied by a 'safety factor') to the allowable tensile stress of float glass.	By multiplying the load resistance of a float glass element by some load duration-dependent glass type factor.‡

* The generally used parameter set does, however, not directly reflect test data, see Section 4.3.1.
† In contrast to usual characteristic resistance values, this one is not a 'real' material parameter. It depends on the geometry, the surface condition, the environment and the loading of the specimens. The term 'characteristic value' is therefore somewhat misleading, which is why it is put in inverted commas.
‡ In CAN/CGSB 12.20-M89 [60], the glass type factor (called 'heat treatment factor') does not depend on the load duration. The load duration factor, however, is glass type dependent, which comes to the same thing.

Load duration effects

Time-dependent effects related to loads are commonly referred to as '*load duration effects*' or '*duration-of-load effects*'.[15]

[15] Strictly speaking, the term is not very accurate because it implies constant loads. In the more general case of time-variant actions, 'action history effects' would be more appropriate. As 'load duration effect' represents

All design methods presented in this chapter are implicitly or explicitly based on the assumption that crack growth and with it the probability of failure of a crack or an entire glass element can be modelled using the *risk integral*, also known as *Brown's integral* (see Section 3.3.4).

Lifetime prediction based on the risk integral implies that the failure probability can be described accurately by accounting for the crack growth (damage accumulation) during the lifetime only while neglecting the influence of the initial surface condition and of the momentary load. These simplifying assumptions have important consequences. The momentary failure probability at a point in time is independent of the momentary load at that time (which is not the case in reality). Furthermore, the material resistance obtained from an equivalent stress based model converges on infinity for very short loading times or very slow sub-critical crack growth. This is clearly inaccurate. It should converge on the inert strength, which is an upper resistance limit (cf. Section 3.3.3).

These issues were among the reasons that led to confusion and a general lack of confidence in advanced glass design methods. Together with the high variance in experimentally determined parameters, they are also the main reasons for various researchers to claim that the glass failure prediction model and any Weibull distribution based design approach are fundamentally flawed and unrealistic (e. g. [58, 59, 287, 288]).

It can be shown that the risk integral is a good approximation if the crack depth at failure is substantially greater than the initial crack depth [189]. It is, therefore, suitable for structural design calculations. With very high loading rates or low crack velocities, i. e. when little sub-critical crack growth occurs, the simplified approach grossly underestimates the probability of failure and leads to unrealistic and unsafe results (resistance above the inert strength). For these cases, which include the interpretation of experiments, the generalized formulation, which is presented in Section 3.3.4, must be used.

All design methods account for the load duration effect by a factor that depends on the load duration and sometimes the residual stress *only*. In European methods, this factor is applied to the allowable maximum or equivalent in-plane principal stress. In GFPM-based methods, it is applied to the allowable lateral load. This is again a simplifying assumption. In reality, the load duration effect additionally depends on many other factors, including the action history and the element's geometry. For guidance on exact calculations, see Chapter 6.

Residual stress

Several methods include the effect of residual stresses within the resistance of the glass. It is, however, crucial to distinguish residual stress clearly from inherent strength . Only decompressed parts of a glass element's surface are subjected to sub-critical crack growth and its consequences. Furthermore, the uncertainties, and consequently the partial factors, are different for residual stress and inherent strength.

Design methods accounting for residual stress explicitly superimpose the residual stress on the inherent strength. This assumes that the inherent strength is not affected by heat treatment. There is evidence showing that the tempering process actually causes a certain amount of 'crack healing' [41, 192]; this assumption can thus be considered safe (conservative) for design.

commonly accepted terminology and is widely used in academic publications, the term is nevertheless used in the present document.

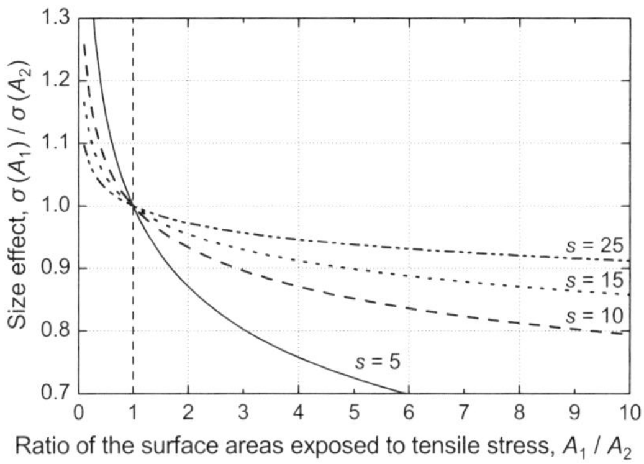

Figure 4.1:
The size effect's dependence on the Weibull shape factor or first surface flaw parameter (using Equation (4.34)).

Size effect

As a direct consequence of the use of Weibull statistics (see below), the resistance of glass elements depends on their surface areas in all design methods. As only tensile stress can cause glass failure, the size depends not on the total, but on the decompressed, surface area. For given geometry and support conditions, the latter depends in general[16] on the load intensity and is therefore time-variant. Taking this aspect accurately into account is complex (see Section 3.3). European design methods define the size factor based on the total surface area, which makes it load independent. US and Canadian standards multiply the load resistance of annealed glass elements by a factor. As the entire surface of an annealed glass plate is immediately decompressed on loading, the two approaches yield the same result. The size effect can be expressed as

$$\frac{\sigma(A_1)}{\sigma(A_2)} = \left(\frac{A_2}{A_1}\right)^{1/s} \tag{4.34}$$

where $\sigma(A_1)$ and $\sigma(A_2)$ are the tensile strengths of structural members with surface areas A_1 and A_2, respectively, exposed to tensile stress. In European methods, s is the shape parameter β of the tensile fracture strength distribution. In GFPM-based methods, s is the surface flaw parameter $\tilde{m}$.[17]. *Figure 4.1* shows that the size effect is quite significant for the ASTM E 1300 value of $\tilde{m} = 7$, while it becomes almost negligible (for realistic panel sizes) for the value $\beta = 25$ that is generally used in European design methods.

Two issues with the size effect are rather problematic: firstly, the exponent s differs much between design methods. Secondly, while the size effect in as-received glass is verified by experimental evidence, the same cannot be said for weathered glass. Calderone [59], for instance, found little or no relationship between the total surface area and the breakage stress or between the most stressed panel area and the breakage stress. This issue is further discussed in Chapter 6.

Weibull statistics

According to the model in Section 3.3, the strength of as-received or homogeneously damaged glass specimens should follow a two-parameter Weibull distribution. This is

[16] In some special but frequent cases such as annealed glass plates exposed to uniform lateral load, the whole surface is decompressed at all non-zero load intensities.

[17] The two parameters $\tilde{m}$ and β are not identical. It is $\beta = \tilde{m}(n+1)/n$, so typically $\beta = 17/16 \cdot \tilde{m}$ [189]

indeed the case with experimental data obtained at *inert* conditions. The goodness-of-fit of the results of common *ambient* strength tests to the Weibull distribution, however, is often poor. Several researchers have, therefore, proposed using log-normal or normal distributions to represent glass strength. This is problematic because the Weibull distribution results from fundamental hypotheses of the model in Section 3.3. The use of a non-Weibull distribution type would require an alternative glass strength model.

In contrast to inert strength data, strength data obtained at ambient conditions do not solely represent a glass specimen's surface condition, but are additionally influenced by sub-critical crack growth. Haldimann [189] showed that the goodness-of-fit of ambient strength data to the Weibull distribution decreases as time to failure increases, which means as the influence of sub-critical crack growth increases. The poor fit of ambient strength data to the Weibull distribution can, therefore, be explained by the wide variability of the crack velocity parameters (cf. Section 3.2).

In conclusion, Weibull statistics are in principle well suited to describe the strength of glass. The problems encountered are not related to the statistical model itself. Instead, they are caused by the variability of the crack growth parameters and by the fact that the damage on glass elements in in-service conditions is often not uniform and homogeneous. This issue is further discussed in Chapter 6.

Biaxial stress fields

The GFPM-based standards use the biaxial stress correction factor proposed by the GFPM. It depends on the principal stress ratio (which is, in general, load intensity dependent) and on the surface flaw parameter $\widetilde{m}$. Though not explicitly stated, only the fully developed principal stress ratio is used for the resistance graphs and the testing procedure. Based on this ratio, a single biaxial stress correction factor for each point on the surface is calculated and assumed to be valid for all load intensities.

European glass design methods generally assume all cracks to be oriented perpendicularly to the major principal stress. This is equivalent to assuming an equibiaxial stress field. Although this assumption is conservative (safe) for design, it is *not* conservative when deriving glass strength data from tests (see Section 6.4).

4.6 Conclusion and Outlook

Current widely used design methods suffer from notable shortcomings. They are, for instance, not applicable to general conditions, but are limited to special cases like rectangular plates, uniform lateral loads, constant loads, time-independent stress distributions and the like. Some model parameters combine several physical aspects, so that they depend on the experimental setup used for their determination. The condition of the glass surface is not represented by user-modifiable parameters, but is embedded implicitly. Finally, the design methods contain inconsistencies and different models yield differing results.

The following two chapters present recent findings that endeavour to redress these issues. Chapter 5 explains ways to design structural glass elements for compressive in-plane loads and stability problems. Chapter 6 explains how the design methods discussed in the present chapter can be generalized and their scope of application can be extended based on the considerations in Chapter 3.

Chapter

5 Design for Compressive In-plane Loads and Stability Problems

5.1 Introduction

The compressive strength of glass is significantly higher than its tensile strength [80, 169]. Experimental studies [242] demonstrated that it is possible to utilize the enormous compressive in-plane load carrying capacity of glass panels. This opens up new applications of glass panels in structures such as columns, transparent walls, beams, for fins to stiffen façade elements, for shear panels, and for applications where the glass is used in a similar way to steel, aluminum or timber [233, 339]. Owing to the high slenderness of structural glass elements made of thin glass plates, they are unstable and tend to fail. Every in-plane loaded glass element must, therefore, be checked against stability failure. Several established design methods exist for common structural materials (i. e. steel, timber), but these methods cannot be applied directly to glass, because the influence of production tolerances (thickness, variation in panel size) of the initial imperfections, of the brittle behaviour, and of the viscoelastic behaviour of laminated glass interlayers have to be specifically considered for glass. A substantial amount of fundamental research has been carried out in the past few years to investigate the stability behaviour of structural glass elements. Nevertheless, results are not yet implemented in existing design standards. Column buckling of glass elements was studied by Kutterer [234], Luible [243, 247], and Overend [268]. Fundamental research on lateral torsional buckling of glass beams was done by Belis [36], Holberndt [240], Kasper [224] and Luible [243, 245]. Research on glass plate buckling is a relatively new research field. First experimental and analytical studies were carried out by Englhardt [162], Luible [243] and Wellershoff [338, 339].

In the past, stability problems were described with bifurcation buckling models based on linear elastic stability theory. The bifurcation buckling theory assumes that a geometrically perfect elastic structural member that is subjected to an increasing load fails suddenly when a critical load is reached. This critical load depends only on the geometry, the loading conditions and the flexural stiffness of the element and may be determined by mathematical models (i. e. [325]) or by numerical approaches such as finite element analysis (FEA). Bifurcation buckling models are generally unable to describe the buckling strength of real structural elements with initial out of straightness and non-linear material behaviour (such as in the case of steel or aluminium). The critical buckling load represents an upper strength limit for column buckling and lateral torsional buckling. In the case of plate buckling, loads higher than the critical buckling load may be applied because of the so-called post buckling behaviour of plates, which is a consequence of membrane effects.

Because of geometric imperfections (i. e. initial deformations w_0 and v_0, see *Figure 5.1*), the load carrying behaviour of stability-critical structural elements is characterized by deformations, even for very small loads. A further increase of the load leads to a non-linear increase of the deformations until the strength of the material or a deformation limit is reached. Therefore, bifurcation buckling models are not satisfactory for design. Nevertheless, they are of great importance because the critical buckling loads calculated this way are often used as reference values for design aids like buckling curves. More realistic models and approaches, such as second order models and non-linear numerical buckling analysis, are required to describe this load carrying behaviour. At this stage it is important to note that non-linear material behaviour does not need to be taken into account because of the ideally elastic behaviour of glass.

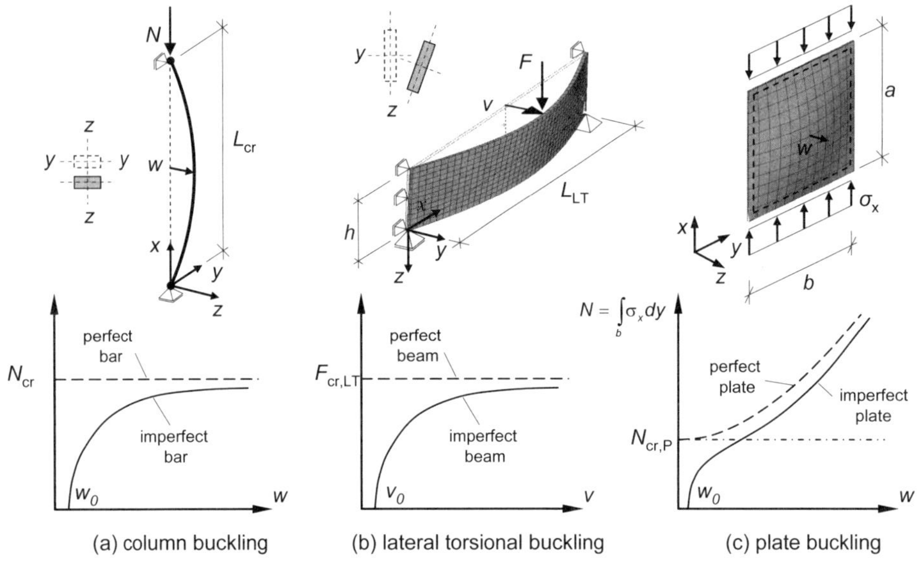

Figure 5.1: Fundamental stability problems and load carrying behaviour: (a) column buckling, (b) lateral torsional buckling, (c) plate buckling.

5.2 Parameters having an influence on the buckling behaviour

The buckling behaviour of structural glass elements is mainly influenced by

- production tolerances (i. e. glass thickness, geometrical imperfections),
- the initial deformation,

- the interlayer material used in laminated glass,
- the linear elastic material behaviour without plastic deformability or strain hardening effect as for steel,
- the fracture strength, which depends on the inherent glass strength and on the residual stress (see Chapter 3),
- the boundary conditions (i. e. type of fixing, silicone joints and gaskets).

Some of these influencing aspects have recently been studied [243, 298]. The main findings are summarized in the following sections.

5.2.1 Glass thickness

Glass manufacturers try to save material in making the best use of the thickness tolerances specified by the codes [155]. The real glass thickness t is generally less than the nominal value, therefore reducing the moment of inertia of the cross section and, thus the buckling strength. Measurements showed that glass thickness values follow a normal distribution. The 5% fractile value corresponds to 97.6% of the nominal glass thickness [243].

5.2.2 Initial deformation

The initial geometric deformation (w_0 and v_0, see *Figure 5.1*) is mainly caused by the tempering process. The measurements confirmed that non tempered annealed flat glass has a very low initial deformation ($< L/2500$), while heat-strengthened glass (HSG) and fully tempered glass (FTG) can have a sinusoidal initial deformation up to $L/300$. Laminated glass showed the same results as monolithic glass. The measured values fitted well to a normal distribution with a 95% fractile value of $L/386$. Maximum initial deformations, however, depend strongly on the quality of the tempering furnace and can therefore vary among manufacturers [243].

5.2.3 Interlayer material behaviour in laminated glass

Different interlayer materials such as polyvinyl butyral (PVB) or DuPont's SentryGlass Plus, which are used in laminated glass, are discussed in Section 1.3.3.

The viscoelastic material behaviour of the PVB interlayer, for example, leads to a time and temperature dependent bending behaviour of laminated glass. The resistance against stability failure is significantly higher under short-term loads than it is under long term loads. As calculations with viscoelastic models are complicated, simplified approaches may be used [234, 243, 338] instead, allowing for a calculation with an equivalent elastic material.

The design approaches presented in this chapter are developed on laminated glass with PVB interlayer. As long as the interlayer may be simplified as a linear elastic material, they may be applied to other interlayer materials as well.

5.2.4 Boundary conditions and glass fixings

The support conditions and the way the load is applied on a structural glass member may have a positive effect, for instance when the rotation of an edge is partially restrained by its fixing, silicone joints or gaskets. Such support conditions tend to reduce the effective buckling length and thus increase the buckling resistance. However, if these effects are taken into account in the design process, an adequate model has to be used and the joint quality has to be monitored during the lifetime.

A negative effect may occur if the glass fixing creates an eccentric load introduction, thus resulting in additional bending moments which reduce the buckling resistance.

5.3 Column buckling

5.3.1 Modelling

The load carrying behaviour of *monolithic* glass can be described using the second order differential equation for a bar with length L_{cr} and pinned ends, with an initial sinusoidal deformation w_0 and an axial compression load N, which is applied with an eccentricity e (*Figure 5.2*).

$$EI\frac{d^2w(x)}{dx^2} + N\left[w_0 \sin\frac{\pi x}{L_{cr}} + e + w(x)\right] = 0 \tag{5.1}$$

The elastic critical buckling load is

$$N_{cr} = \frac{\pi^2 EI}{L_{cr}^2} \tag{5.2}$$

and the maximum deflection w at midspan considering second order effects is

$$w_{max} = \frac{e}{\cos\left(L_{cr}/2\sqrt{N/N_{cr}}\right)} + \frac{w_0}{1 - N/N_{cr}} \,. \tag{5.3}$$

This yields a maximum surface stress

$$\sigma_{max} = \frac{N}{A} \pm \frac{M}{W} = \frac{N}{A} \pm \frac{N}{W}(w_{max} + w_0 + e) \,, \tag{5.4}$$

where N is the applied force, A is the sectional area and W is the elastic section modulus.

In *laminated glass* the interlayer material behaves like a shear connection between the glass panes. Simplistically, several interlayer materials may be considered as elastic

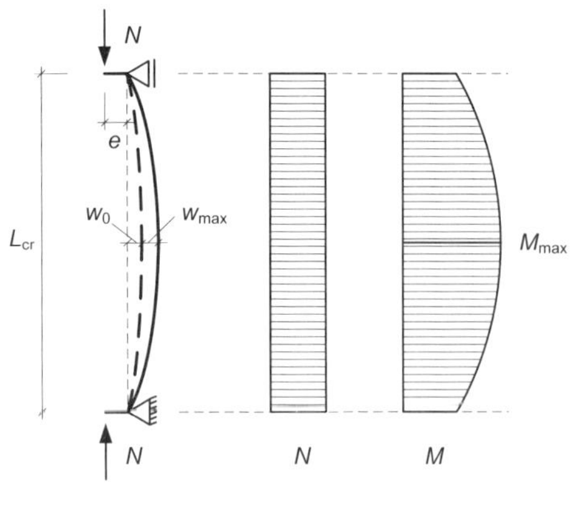

Figure 5.2:
Column buckling model.

materials with a constant shear stiffness for a given temperature and load duration. The load carrying behaviour can then be described using elastic 'sandwich' theory [322, 351]. The critical buckling load of a laminated glass with two or three glass panes and symmetrical layout (*Figure 5.3*) is given in [243] as:

$$N_{cr} = \frac{\pi^2(1+\alpha+\pi^2\alpha\beta)}{1+\pi^2\beta}\frac{EI_S}{L_{cr}^2} \tag{5.5}$$

In the case of two glass panes, it is

$$\alpha = \frac{I_1+I_2}{I_S} \quad ; \quad \beta = \frac{t_{int}}{G_{int}b(z_1+z_2)^2}\frac{EI_S}{L_{cr}{}^2} \quad ; \quad I_S = b(t_1z_1^2+t_2z_2^2)\,, \tag{5.6}$$

and in the case of three glass panes, it is

$$\alpha = \frac{2I_1+I_2}{I_S} \quad ; \quad \beta = \frac{t_{int}}{2G_{int}bz_1^2}\frac{EI_S}{L_{cr}{}^2} \quad ; \quad I_S = 2bt_1z_1^2\,. \tag{5.7}$$

L_{cr} is the buckling length, b is the width of the cross section, G_{int} is the shear modulus of the interlayer material and $I_i = bt_i^3/12$ is the moment of inertia of the pane i.

A simplified approach for calculating the deflection and the maximum bending stresses of a laminated glass consists in employing Equations (5.4) and (5.3) with the following equivalent thickness [243]:

$$t_{eff} = \sqrt[3]{\frac{12I_S(1+\alpha+\pi^2\alpha\beta)}{b(1+\pi^2\beta)}} \tag{5.8}$$

A and W in Equation (5.4) are $A = b\sum t_i$ and $W = bt_{eff}^2/6$.

It is assumed that the glass pane's rotation is not restrained at either extremity and that the load is applied axially, i. e. there are no lateral loads.

Kutterer [234] developed an analytical second order model based on sandwich theory [322] for the analysis of the buckling behaviour of laminated glass elements under an axially applied force. The model accounts for creep effects of the PVB interlayer. The lateral displacement and the maximum stresses may be calculated as a function of temperature and load duration. For a given axial load the model is able to predict a critical time at which time delayed buckling will start.

Analytical models are generally limited to simple structural systems and certain boundary conditions. Numerical finite element models are more flexible and powerful. They have the advantage that the interlayer may either be represented by elastic or viscoelastic elements based on existing material data [334]. Furthermore, arbitrary boundary conditions (e. g. restraints due to the load introduction or intermediate supports) may be easily incorporated (see Section 2.3.2).

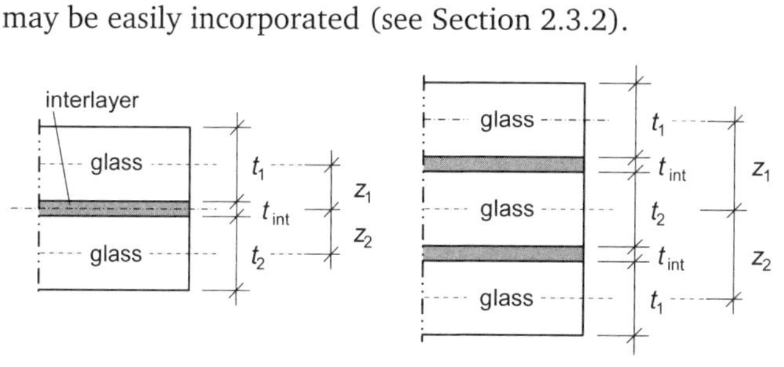

Figure 5.3:
Cross section of a laminated glass with two (left) and three (right) glass panes

5.3.2 Load carrying behaviour

The column buckling behaviour of glass elements made of monolithic and laminated glass with PVB interlayer was studied experimentally and compared to analytical and numerical models by Luible [243] and Kutterer [234]. The models in Section 5.3.1 are suitable to describe the load carrying behaviour of glass elements with imperfections in compression. In contrast to monolithic glass, the buckling behaviour of laminated glass depends on load duration and temperature because of the viscoelastic behaviour of the PVB interlayer [244]. Several other parameters have an influence on the column buckling behaviour:

- The glass thickness t, the initial deformation w_0 and the load eccentricity e have the most important influence on the buckling strength. The real glass thickness rather than the nominal glass thickness has to be taken into account (see Section 5.2.1). The buckling strength that results from the real thickness may be up to 11.7% less than the buckling strength that is obtained based on the nominal thickness.
- Because of the high compressive strength of glass, the failure origin of glass element in compression is always on the tension surface for the panel dimensions commonly used in buildings ($L > 300\,\text{mm}$, $t < 19\,\text{mm}$) and initial deformations as explained in Section 5.2.2. The buckling strength of glass is, therefore, limited by the maximum tensile strength [243].
- Experimental studies demonstrated that the failure origin is mainly in areas that have a low tensile strength. In the case of annealed glass, this is the glass edge (most severe surface damage). In the case of tempered glass, this is close to the glass edge, where the residual compressive surface stress reaches a minimum (see Section 3.6.4).
- The load carrying capacity of tempered glass is mainly influenced by the residual compressive surface stress rather than by the inherent glass strength. This effect is caused by the non-linear relationship between the applied compressive load and the tensile stress on the glass surface (*Figure 5.4*). As a simplified approach for column buckling design, the inherent strength may be neglected and the buckling strength can be determined from the residual stress only.
- The composite action caused by the viscoelastic PVB interlayer in laminated glass increases the buckling strength. Unfortunately, creep effects in the PVB make the buckling strength depend on time and temperature. For low temperatures and short-term loading, the buckling strength can almost reach the buckling strength of a monolithic cross section of the same thickness. For long-term loads (e. g. dead

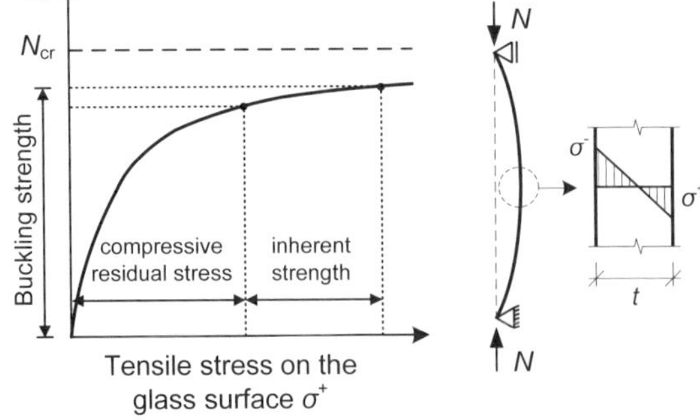

Figure 5.4: The influence of the residual stress and of the inherent strength on the buckling strength.

load) or high temperatures (> 50 °C) the composite action provided by the PVB is insignificant and the load carrying behaviour is similar to independent glass panes without PVB. Therefore the lower limit of the buckling strength of a laminated glass may be determined by ignoring the composite action ($G_{PVB} = 0$). From a safety point of view, a shear interaction may only be taken into account for short-term loads like wind or impact loads and for temperatures < 25 °C.

5.3.3 Structural design

Compressive members such as steel columns are generally designed using column curves. This approach can be applied to compressive glass elements as well. In steel construction, column curves are based on a slenderness ratio [143]. This allows the same curve to be used for the design of members with different steel grades. However, in contrast to steel, the slenderness ratio for glass must be based on the maximum tensile strength, as the compressive strength does not limit the buckling strength. The application of column curves for column buckling of glass elements has been discussed in [244]. It is shown that as a simple approach, the maximum tensile stress in a compressed glass member can be determined by means of elastic second order equations (Equation (5.4)). The column buckling capacity of a glass element is adequate if

$$f_{Sd} \leq f_{Rd} \tag{5.9}$$

where f_{Sd} is the design value of the maximum tensile stress and f_{Rd} the design value of the maximum tensile strength.

A reduced glass thickness and a reasonable assumption of the initial deformation has to be considered in the second order analysis (Section 5.2). Due to the non-linear relation between applied loads and resulting bending stresses, the maximum bending stress has to be determined with factored load and superposition of stresses resulting from different loads is not possible. Simplistically the tensile strength may be assumed to be equal to the residual stress (see Section 5.3.2).

This approach applies to laminated glass as well. Generally it is advantageous to take the composite behaviour of the interlayer into account. For PVB interlayers it is recommended to consider a composite behaviour only for short-term loads such as wind loads. Simplistically, the sandwich cross section may be replaced by an effective monolithic cross section with an effective thickness t_{eff} given by Equation (5.8). Maximum stresses may be calculated with Equation (5.4) or numerical models.

5.3.4 Intermediate lateral supports

The buckling strength of a member with monolithic cross section is directly proportional to the square of the buckling length. Reducing the effective buckling length by means of an additional intermediate lateral support will increase the buckling strength by a factor of 4 (*Figure 5.5*).

$$\frac{N_{cr,2}}{N_{cr,1}} = \frac{L^2}{\left(\frac{L}{2}\right)^2} \Rightarrow N_{cr,2} = 4N_{cr,1} \tag{5.10}$$

This assumption is not valid for laminated glass. Decreasing the effective buckling length of a laminated glass by intermediate lateral supports increases the buckling strength but on the other hand it also decreases the lateral bending stiffness, which is a function of the interlayer shear modulus and the effective shear length. The shear length may be conservatively assumed as the distance between the supports. For a more realistic analysis it is recommended to use suitable finite element models where the entire member is modelled [243]. $N_{cr,3}$ may be used as a conservative approach to determine the buckling strength of $N_{cr,2}$ (*Figure 5.5*). For laminated glass it may be assumed that

$$N_{cr,3} \leqslant N_{cr,2} \leqslant 4N_{cr,1} \tag{5.11}$$

.

Figure 5.5: Influence of intermediate lateral supports on the critical buckling load

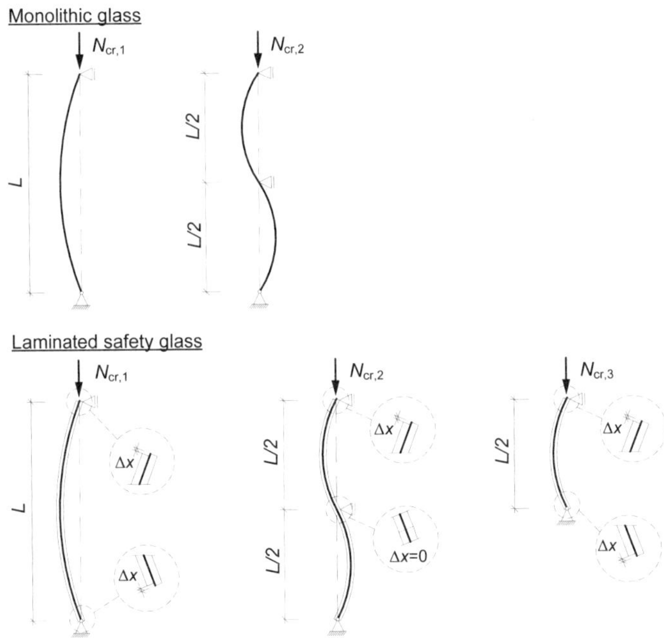

5.3.5 Influence of the load introduction

The laminated glass models are based on the assumption that the shear deformation between the glass panes at the extremities is free and that the load is applied symmetrically on the cross section. In practice, the glass edges of a laminated glass made of HSG or FTG are not flush and loads are generally applied through intermediate materials such as neoprene, injected mortar, high strength plastics or aluminium. The partial shear restraint due to these load introduction materials leads to a stiffer load carrying behaviour than assumed in the model. Uneven glass edges result in an asymmetric load distribution on the glass panes, which creates additional bending moments. Such bending moments may be determined, for example, with the model presented in [243], which takes asymmetric thickness of the load introduction material into account. If the influence of the glass edges is critical, detailed finite element models are recommended for design.

5.4 Lateral torsional buckling

Lateral torsional buckling is a limit state of structural stability, where a beam is subjected to bending. The typical structural deformation is a combination of lateral deflection and twisting (see *Figure 5.1*). In glass structures, lateral torsional buckling may, for instance, occur in glass beams or glass fins used as lateral stiffeners in façades.

5.4.1 Modelling

The critical torsional buckling moment (bifurcation buckling) of a beam with a rectangular cross section can generally be determined by

$$M_{\mathrm{cr,LT}} = C_1 \frac{\pi^2 E I_z}{L_{\mathrm{LT}}^2} \left[\sqrt{C_2 z_a + \frac{GK\, L_{\mathrm{LT}}^2}{\pi^2 E I_z}} + C_2 z_a \right] \tag{5.12}$$

where E is Young's modulus, I_z is the moment of inertia about the z-axis, G is the shear modulus, K is the torsion constant, and L_{LT} is the unrestrained beam length. The factors C_1 and C_2 take into account different bending moments (*Table 5.1*) and z_a is the distance between the centre of gravity and the point where the load is applied. Due to the rectangular cross-section of monolithic and laminated glass beams, warping torsion may be neglected in practice. In [240] a slightly different formula, based on [291], that accounts for warping torsion effects is proposed for the calculation of the critical buckling moment.

The critical lateral torsional buckling moment of laminated glass may be calculated using Equation (5.12), where the lateral bending stiffness EI_z and the torsional stiffness GK are replaced by an equivalent stiffness, $EI_{z,\mathrm{eff}}$ and GK_{eff}. Both stiffnesses are based on sandwich theory [322, 351] in order to take into account the composite action of the interlayer in laminated glass [243]. The equivalent bending stiffness for laminated glass with two or three glass panes is:

$$EI_{z,\mathrm{eff}} = EI_s \left(\frac{\alpha\beta\pi^2 + \alpha + 1}{1 + \pi^2\beta} \right) \tag{5.13}$$

In the case of two glass panes, it is

$$\alpha = \frac{I_1 + I_2}{I_S} \quad ; \quad \beta = \frac{t_{\mathrm{int}}}{G_{\mathrm{int}} h (z_1 + z_2)^2} \frac{EI_S}{L_{\mathrm{LT}}{}^2} \quad ; \quad I_S = h(t_1 z_1^2 + t_2 z_2^2)\,, \tag{5.14}$$

and in the case of three glass panes, it is

$$\alpha = \frac{2I_1 + I_2}{I_S} \quad ; \quad \beta = \frac{t_{\mathrm{int}}}{2G_{\mathrm{int}} h z_1^2} \frac{EI_S}{L_{\mathrm{LT}}{}^2} \quad ; \quad I_S = 2ht_1 z_1^2\,. \tag{5.15}$$

Table 5.1: Lateral torsional buckling factors C_1 and C_2.

Bending moment	C_1	C_2
Constant	1.0	–
Linear (zero at mid span)	2.7	–
Parabolic (zero at both extremities)	1.13	0.46
Triangular (zero at both extremities and maximum at mid span)	1.36	0.55

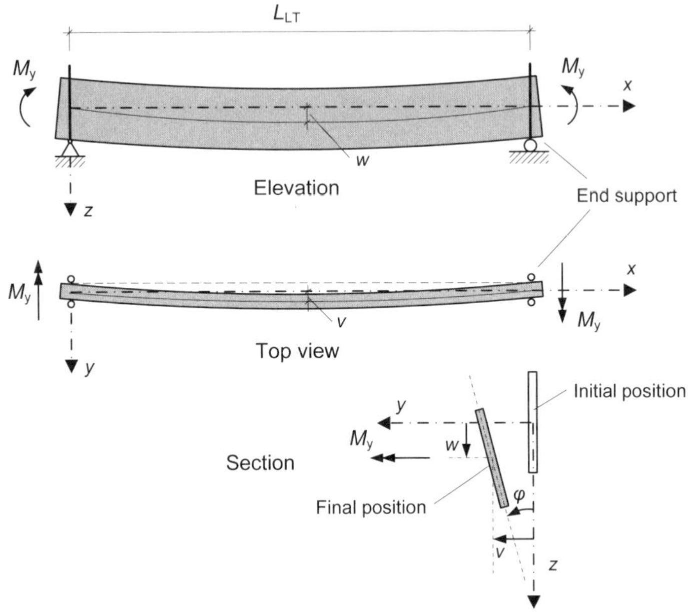

Figure 5.6: Lateral torsional buckling model with bending moments applied at both extremities.

The variables t_1, t_2, t_{int}, and z_1 are explained in *Figure 5.3*; h is the beam height, E is Young's modulus, I_i is the moment of inertia of glass pane i, G_{int} is the shear modulus of the interlayer and L_{LT} is the buckling length. The equivalent torsional stiffness GK_{eff} for laminated glass is

$$GK_{eff} = \begin{cases} GK_{glass1} + GK_{glass2} + GK_{comp} & \text{two glass panes} \\ GK_{glass1} + GK_{glass2} + GK_{glass3} + GK_{comp} & \text{three glass panes} \end{cases} \tag{5.16}$$

with

$$GK_{comp} = GI_{S,comp}\left(1 - \frac{2}{\lambda h}\tanh\frac{\lambda h}{2}\right) \tag{5.17}$$

$$I_{S,comp} = \begin{cases} 4\left(\frac{t_1+t_2}{2} + t_{int}\right)^2 \frac{t_1 t_2}{t_1+t_2} h & \text{two glass panes} \\ 2(t_2 + 2t_{int} + t_1)^2 t_1 h & \text{three glass panes} \end{cases} \tag{5.18}$$

The aforementioned formulas are sufficiently accurate for the determination of the critical buckling load of glass beams. In order to describe the non-linear lateral torsional buckling behaviour of an initially imperfect glass beam, analytical models have been developed for the basic structural system of a simple beam with uniformly applied load, constant bending moments and concentrated load at mid span [224]. Those analytical models are limited to small deformations and bending moments $M < 0.8M_{cr,LT}$. Numerical models such as finite element models are generally more powerful. They enable arbitrary boundary conditions and structural systems as well as the non-linear behaviour with significant lateral deformations to be modelled [243]. For monolithic glass, shell elements are sufficient. For laminated glass, different glass panes and interlayers have to be taken into account. A simple approach, in order to reduce the size of laminated glass models, is to model the glass panes with shell elements and the interlayer with volume elements. Shell elements and volume elements are either tied together with coupled nodes or modelled with identical nodes for the shell and volume elements. In the latter case, the shell elements must be defined with an offset (*Figure 5.7*). A lot of interlayer materials may simply be modelled as an elastic material with an appropriate shear stiffness or as a viscoelastic material based on existing material models [334].

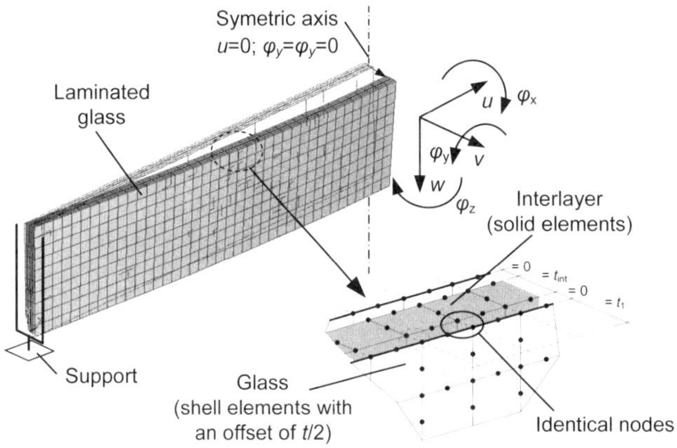

Figure 5.7:
Typical lateral torsional buckling model with shell elements for glass and volume elements for the interlayer.

A typical numerical stability analysis comprises the following steps:

- Creating the model with appropriate elements and material definition.
- Application of the boundary conditions such as vertical and lateral supports and loads. In case of laminated glass, applied conditions have to allow for free rotation and shear deformation of the glass.
- Start of the simulation with a modal analysis of the system. The resulting eigenvalue corresponds to the critical buckling load, the resulting first eigenvector corresponds to the first critical buckling shape of the initial deformation.
- Application of the initial deformation using a scaled shape of the first eigenform of the system.
- Non-linear analysis on this 'imperfect' system.
- Post-processing in order to identify the maximum deflection and principal surface stress.

5.4.2 Load carrying behaviour

The load carrying behaviour of glass beams made of monolithic and laminated glass with PVB interlayer was studied with tests and numerical models in [36, 224, 243, 247]. It turned out that even under small loads and initial imperfections, the top chord of the beams tend to deform laterally. The relation between applied load and lateral beam deflection is non-linear. Similar to column buckling, the monolithic glass shows an elastic behaviour (*Figure 5.1*), while the load carrying behaviour of laminated glass is characterized by the viscoelastic material PVB. Temperature and load duration have a major influence on the buckling strength of laminated glass. The typical load carrying behaviour is explained in *Figure 5.8*:

- A laminated glass beam that is subjected to a constant force F shows a disproportional increase of the lateral deformation due to the creeping of the PVB (*Figure 5.8*). If the applied force F is less than the buckling strength of the glass panes without taking the PVB interlayer into account ($G_{\mathrm{PVB}} = 0$), the lateral deflections will converge. For load durations $t \to \infty$, the lateral deflections converge on the value of a glass beam with independent panes. For loads higher than the buckling strength of the glass without taking the PVB interlayer into account, creeping effects lead to a disproportional non-converging increase of the lateral deflection until the glass breaks.
- A laminated glass beam subjected to a linearly increasing displacement u' (e. g. in a lateral torsional buckling test) has a load versus lateral deflection behaviour as shown in *Figure 5.8*. The applied force increases up to a critical maximum until creep effects become important and the deflection increases disproportionately. Higher displacement rates lead to higher peak values. The force converges on the buckling strength of independent glass panes ($G_{\mathrm{PVB}} = 0$).

The glass beam fails and the buckling resistance is attained when the maximum tensile strength on the glass surface is exceeded. In general this buckling resistance occurs at load levels lower than the critical buckling load calculated with the bifurcation buckling model. Exceptions are very slender glass beams with a long span and a high ratio of beam height to glass thickness. Such beams show high lateral deflections and a warping deformation of the cross section. This deformation causes a load carrying behaviour which is closer to the behaviour of shell structures than of beams and the buckling resistance is higher than the critical buckling load [243].

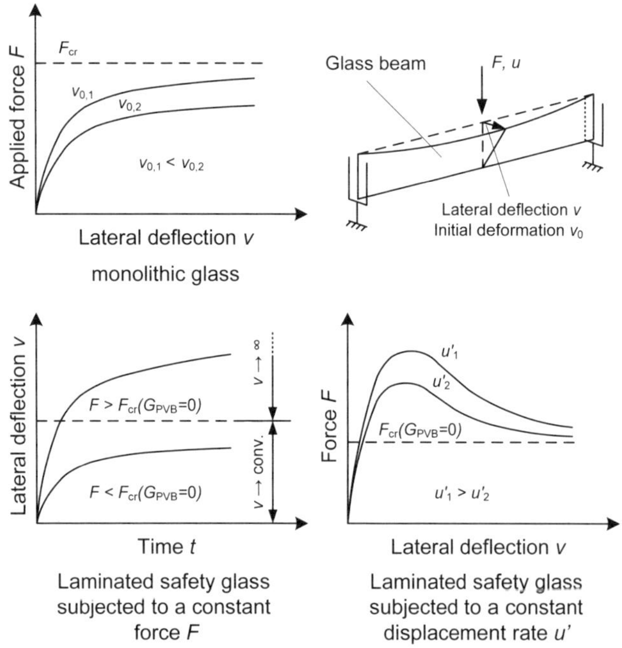

Figure 5.8: Typical lateral torsional buckling behaviour of monolithic and laminated glass beams.

Because of the high compressive strength, the buckling resistance of glass beams is governed by the tensile glass strength. As the highest tensile stresses occur on the edge, the tensile strength close to the edge is generally critical for buckling failure. Failure origin monitoring during buckling tests on HSG and FTG showed three critical areas where the stresses exceeded the tensile glass strength (*Figure 5.9*), which are:

- the corner of the edge (a);
- the lateral surface in a certain distance from the edge (b);
- the centre of the glass edge (c).

Which failure origin finally causes failure of the beam depends on the residual edge stresses, on surface damages, and on the stress field due to loading.

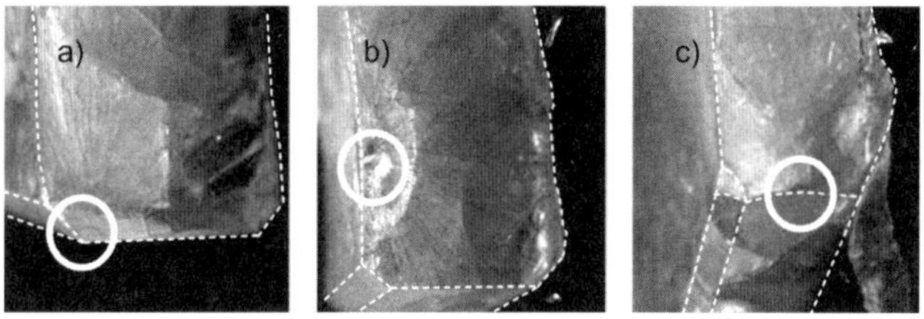

Figure 5.9:
Typical failure origins.

The shear connection by the PVB interlayer has a significant influence on the buckling strength. *Figure 5.10* shows the influence of the PVB shear modulus G_{PVB} on the critical buckling moment $M_{cr,LT}$ and compares it to the case without shear connection ($M_{cr,LT,without\ PVB}$). For realistic values of G_{PVB} (< 5 MPa), $M_{cr,LT}$ is at best 3.2 times $M_{cr,LT,without\ PVB}$. In order to achieve a behaviour that is equivalent to that of a monolithic glass pane, the shear modulus would have to be at least 300 MPa. A significant composite action due to a PVB interlayer may therefore only be taken into account for short-term loads.

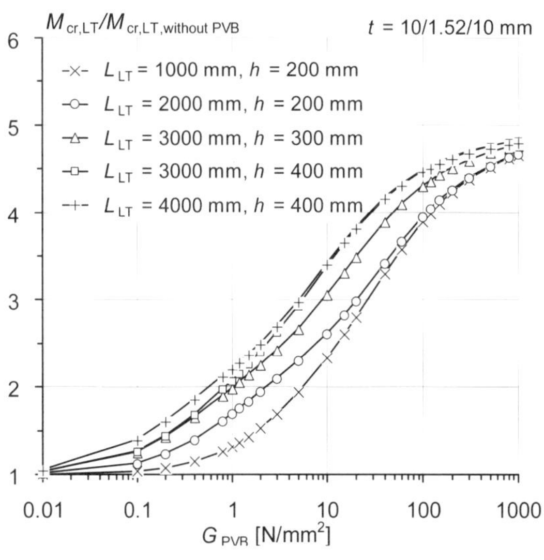

Figure 5.10:
Influence of the shear modulus G_{PVB} on the critical lateral torsional buckling load $M_{cr,LT}$. The curves show the ratio $M_{cr,LT}$ / $M_{cr,LT,without\ PVB}$.

5.4.3 Structural design

Currently there are only few standards and documents dealing with lateral torsional buckling design of glass fins.

The Australian standard AS 1288 [8] provides a design method for glass fins in various structural configurations. The design moment of a glass fin is limited to the critical elastic buckling moment $M_{cr,LT}$ divided by a factor of safety of 1.7. Research has shown [219, 243] that this approach is inappropriate because it does not take the non-linear load carrying behaviour due to imperfections, the glass strength and the PVB interlayer of laminated glass into account.

In [219], a very simple approach for the buckling verification of the unsupported edge of glass beams is given. The method is based on Bleich's concept, where the maximum destabilizing bending moment M_{max} is determined with

$$M_{max} = \frac{Et^3}{6(1+\nu)} \tag{5.19}$$

This equation does not account for the non-linear load carrying behaviour due to imperfections, the glass strength and the PVB interlayer of laminated glass. It should, therefore, only be use for preliminary design.

In general, there are three different approaches for lateral torsional buckling design of glass beams:

- non-linear numerical models (FEA),
- analytical models based on second-order theory,
- buckling curves derived from buckling tests and numerical FEA models.

The lateral torsional buckling resistance of a glass beam may be determined by means of an appropriate finite element model (*Figure 5.7*). Such models allow non-linear effects, initial imperfections and arbitrary boundary conditions to be taken into account.

Analytical models based on second order theory are limited to elementary structural systems and boundary conditions. They are also limited to small deformations and the application of the models in practice is relatively complex.

The buckling design with design curves, such as those adopted in steel and timber design, provide a simple and quick design method. This method is also convenient for glass. Lindner and Holberndt [240] and Luible and Crisinel [246] investigated the application of buckling curves for the lateral torsional buckling design of glass beams. Generally, buckling curves give reduction factors χ_{LT} for design as a function of the slenderness ratio $\overline{\lambda}_{LT}$. In contrast to steel, both parameters are based on the tensile strength of glass. The slenderness ratio is a function of the structural system, the boundary conditions, the loading conditions and the glass type:

$$\overline{\lambda}_{LT} = \sqrt{\frac{\sigma_{Rk}}{\sigma_{cr,LT}}} = \sqrt{\frac{2\sigma_{Rk} I_y}{M_{cr,LT} h}} \tag{5.20}$$

σ_{Rk} is the characteristic tensile strength and $\sigma_{cr,LT}$ is the critical lateral torsional buckling stress. The critical lateral torsional buckling moment $M_{cr,LT}$ may be calculated with Equation (5.12). For the design of laminated glass elements, the equivalent lateral bending stiffness $EI_{z,eff}$ (Equation (5.13)) and the equivalent torsional stiffness GK_{eff}

(Equation (5.16)) may be used. The reduction factor is defined as

$$\chi_{LT} = f\left(\overline{\lambda}_{LT}\right) \tag{5.21}$$

Hence the design value of the bending moment capacity of the glass beam becomes

$$M_{LT,Rd} = \chi_{LT} \cdot \sigma_{Rd} \cdot W_y \tag{5.22}$$

where σ_{Rd} is the design value of the tensile strength and W_y is the section modulus about the strong axis of the beam. For various loading conditions (linear load, concentrated load, constant bending moment), glass geometries, interlayer shear moduli, and initial deformations v_0, reduction factors were simulated using numerical models, compared to experimental test results and plotted in buckling diagrams [240, 243]. An example taken from [243] is shown in *Figure 5.11*. Since there are currently no design methods or codes, these diagrams may serve as a preliminary orientation. The following should be noted:

- Lindner and Holberndt [240] and Luible [243] confirmed that it is possible to define lateral torsional buckling curves for glass based on the tensile glass strength.
- Based on tests and numerical simulations, reduction factors for monolithic glass [240, 243] and for laminated glass [243] were determined (*Figure 5.11*). These authors also discuss possible buckling curves.
- Reduction factors for laminated glass are lower than for monolithic glass.
- As a conservative approach, the buckling curve (c) of the European steel design code [142] might be used for the design of monolithic and laminated glass beams that are subjected to concentrated loads, uniformly distributed loads or constant bending (*Figure 5.11*). Luible [243] showed that all reduction factors found from simulations and laboratory tests are located above this curve.

The lateral torsional buckling verification for glass beams is as follows:

$$M_{LT,Sd} \leq M_{LT,Rd} \tag{5.23}$$

$M_{LT,Sd}$ is the design value of the bending moment due to applied loads.

In practice, additional criteria might have an influence on the lateral torsional buckling of glass beam as well. These are:

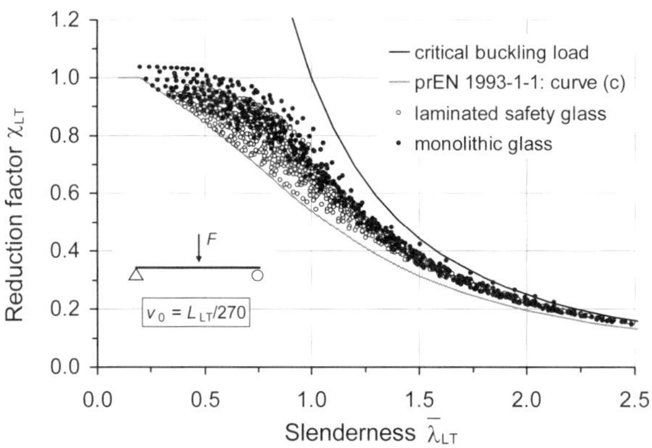

Figure 5.11: Reduction factors χ_{LT} for lateral torsional buckling of a glass beam subjected to a concentrated force at mid-span with $v_0 = L_D/270$.

- Silicone joints or gaskets generally create an additional lateral restraint of the beam top chord and therefore increase the buckling resistance. If the functioning of the silicone joint and gaskets can be guaranteed over the entire design life, they may be considered as additional elastic supports [38].
- The unfavourable influence of details such as supports or restraints on the load carrying behaviour has to be studied carefully during thc dcsign process. Some types of supports (e. g. clamps, point fixings) may lead to local stress concentrations in the glass because of their insufficient rotation capacity. This can be more critical than global buckling.

5.5 Plate buckling

This text has been compiled in collaboration with the following experts:
Dr. Frank WELLERSHOFF

Glass plates subjected to in-plane loads tend to fail because of plate buckling. Today, there are only few built structures where glass plates are subjected to in-plane loads. In future this might change as such structures offer new architectural possibilities in terms of high transparency combined with a high load-carrying capacity [243, 338].

Depending on the type of loading, different structural behaviours have to be distinguished:

- plate subjected to pure compression,
- plate subjected to shear,
- plate subjected to compression and shear,
- plate subjected to in-plane and lateral loads.

The ensuing parts of this section show the results of buckling investigations on glass plates that are simply supported along their edges and subjected to combinations of pure compression and shear. A combined loading of compression, shear and lateral loads is yet to be investigated in glass structures.

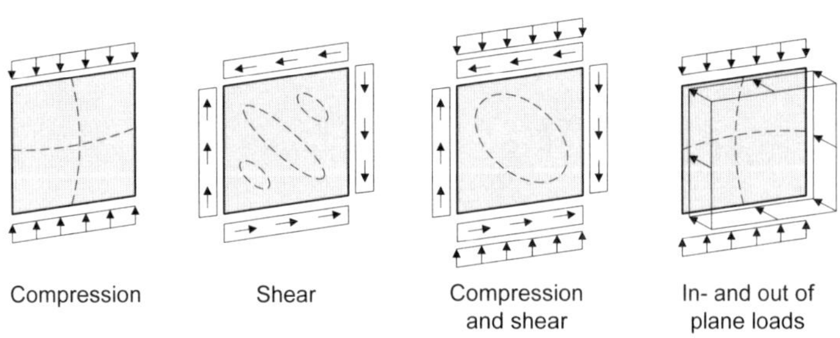

Figure 5.12: Different types of plate buckling.

5.5.1 Modelling

The critical buckling load of a monolithic glass may be calculated with analytical models based on linear elastic bending theory [196]. However, due to post-critical buckling behaviour of plates, the critical buckling loads N_{crit} and τ_{crit} are not criteria for the ultimate strength and thus not suitable for plate buckling design. The critical buckling load overestimates the real buckling strength of compact plates and significantly underestimates the real buckling strength of slender plates. Nevertheless critical buckling load formulas are shown in the subsequent text as they are needed for plate buckling design methods such as buckling curves (see section Section 5.5.3).

The critical buckling load $N_{x,crit}$ (given as force per unit length) of a *monolithic plate* subjected to pure compression (*Figure 5.13*) is

$$N_{x,crit} = \left(\frac{m}{\alpha} + \frac{\alpha}{m}\right)^2 \frac{\pi^2 E\, t}{12\,(1-\nu^2)} \left(\frac{t}{b}\right)^2 \tag{5.24}$$

where $\alpha = a/b$ (*Figure 5.13*), m is number of half sine waves in the x-direction, t is the glass thickness, b is the width of the plate, E is Young's modulus, and ν is Poisson's ratio.

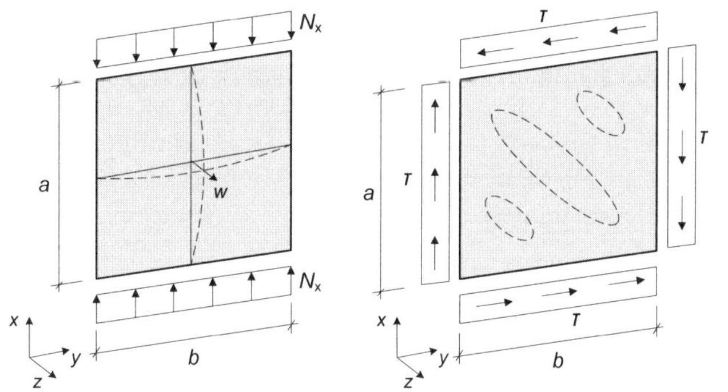

Figure 5.13: Structural system of a four side simply supported plate subjected to pure compression (left) and shear (right).

The critical buckling load $N_{x,crit}$ of a rectangular *laminated glass* plate with two glass panes (*Figure 5.3*) may be determined using linear elastic sandwich theory [351]:

$$N_{x,crit} = \left(\frac{m}{\alpha} + \frac{\alpha}{m}\right)^2 \frac{\pi^2 D}{b^2} \frac{\frac{(D_1+D_2)}{D}\left[\left(\frac{m}{\alpha}\right)^2+1\right] + \frac{Ab^2}{\pi^2 D_s}}{\left[\left(\frac{m}{\alpha}\right)^2+1\right] + \frac{Ab^2}{\pi^2 D_s}} \tag{5.25}$$

where

$$D = D_1 + D_2 + D_s \quad ; \quad D_i = \frac{E t_i^{\,3}}{12(1-\nu^2)} \quad ; \quad D_S = \frac{E t_1 z_1^2 + E t_2 z_2^2}{1-\nu^2} \tag{5.26}$$

$$A = \frac{G_{int}(z_1+z_2)^2}{t_{int}} \tag{5.27}$$

The geometric parameters t_i and z_i are shown in *Figure 5.3*.

The critical buckling load τ_{crit} (given as force per unit length)of a *monolithic plate* subjected to shear (*Figure 5.13*) is

$$\tau_{\text{crit}} = \frac{\pi^2 E t}{12(1-\nu^2)} \left(\frac{t}{b}\right)^2 k_\tau \tag{5.28}$$

with the shear buckling coefficient

$$k_\tau = \begin{cases} 4.00 + 5.34/\alpha^2 & \text{for } \alpha < 1 \\ 5.34 + 4.00/\alpha^2 & \text{for } \alpha \geq 1 \end{cases} \tag{5.29}$$

According to [338], the critical buckling load of *laminated glass* subjected to shear may be determined by

$$\tau_{\text{crit}} = \frac{\pi^2 E t}{12(1-\nu^2)} \left(\frac{t}{b}\right)^2 k_\tau k_{\text{VSG}} \tag{5.30}$$

where k_{VSG} is a correction factor which takes into account the shear stiffness of the interlayer. Values for different shear moduli and geometries are given in [338].

In order to study the load carrying behaviour of a buckled glass plate in a more realistic manner (including post-critical buckling), numerical finite element models (FEM) are recommended. *Figure 5.14* shows a typical finite element model where the glass panes are modelled using shell elements and the interlayer is modelled using volume elements. The two element types are connected using the same nodes. Shell elements are defined with an offset of $t/2$ from the centre of gravity of the glass pane. Load introduction and boundary conditions are applied by means of additional nodes, which are coupled with the element nodes. The edges of the glass panes have to be supported in such a way that the shear deformation is not restrained.

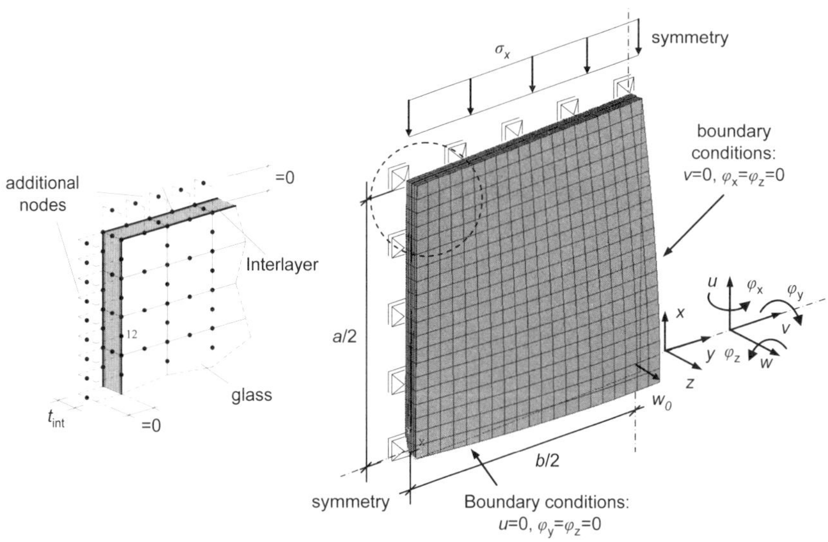

Figure 5.14: Finite element model for plate buckling.

5.5.2 Load carrying behaviour

The plate buckling behaviour of monolithic and laminated glass with PVB interlayer was investigated in [243, 245, 338] with experimental studies and numerical simulations. All tests demonstrated the significant post buckling capacity of glass plates allowing for loads higher than the critical buckling load. The load carrying behaviour depends on whether glass panels are subjected to pure compression or to shear. This is discussed in the following text.

Pure compression

Typical test results of monolithic glass subjected to a uniform pressure [243] are shown in *Figure 5.15*. The ultimate load is twice as high as the critical buckling load. Depending on the slenderness, initial imperfections are less critical for plate buckling when compared to column buckling. The load carrying behaviour depends also on the applied load. Two extreme loading conditions are shown in *Figure 5.15*. In model (a) the load is applied as uniform deformation of the glass edge. In model (b) the load is applied as a uniformly distributed compressive stress. In the test, the load was applied by a steel beam and with an aluminum layer between the glass edge and the steel beam. Therefore, the test results are closer to model (a) in terms of the slope of the curve, but plasticization of the aluminium caused higher deflection w_{max}. Depending on the stiffness of the load introduction materials, the load carrying behaviour lies somewhere between model (a) and model (b). Due to the high elastic energy stored in the glass plate, the breakage is explosive, the fracture patterns for both annealed and heat strengthened glass are untypically very fine, and there is no post-breakage structural capacity.

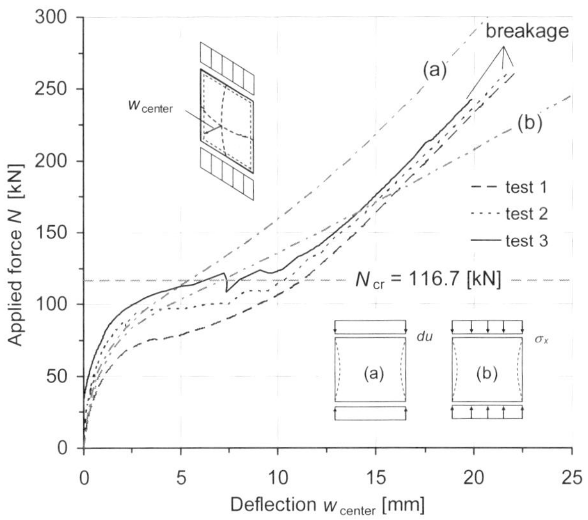

Figure 5.15: Plate buckling test on a monolithic glass (HSG) with dimension 1000 × 1000 mm and a thickness of 8 mm.

In all tests the failure origin occurred on the glass surface and in areas with tensile stress. This means that the tensile strength of the glass surface governs the buckling strength of glass plates. Due to the non-linear behaviour, the location of the maximum in-plane principal stress depends on the load level. At higher loads, this location migrates from plate centre towards the corner (*Figure 5.16*).

Parametric studies showed that the shape of the initial imperfection has an influence on the buckling strength of glass plates [243]. Unlike with column buckling and lateral torsional buckling, the most critical shape for the initial imperfection may not be the first eigenform. Several eigenforms have, therefore, to be checked in order to determine the load carrying capacity. An example is given in *Figure 5.17*. The maximum in-plane principal stress for an applied load of 400 kN is higher for a plate with a double half sine imperfection (EF2) than for a plate with a single half sine imperfection. In other words, EF2 is the most critical imperfection shape for the design of this glass plate if the tensile strength exceeds 60 MPa.

Tests on 1000 mm × 1000 mm laminated glass elements showed only a slight influence of the PVB interlayer on the buckling strength. The comparison with the numerical simulations confirmed that a composite action can be activated, but the shear modulus of the interlayer has to be relatively high to create a noticeable increase in buckling strength [243]. Accordingly, a simple but safe approach for the plate buckling design of laminated glass elements is to neglect the shear stiffness of interlayers. Nevertheless, new and stiffer interlayer materials provide a significantly higher shear modulus, thus substantially improving the plate buckling capacity of laminated glass.

Shear

Wellershoff [338] demonstrated in experimental studies that typical diagonal tension fields can be activated in glass panels (*Figure 5.18*). The edge of the glass panel is glued in a steel frame. The size of the heat strengthened glass splinters is related to the stress concentration at the moment of breakage (Section 3.4). Three areas with higher surface stress concentrations can be identified, which are:

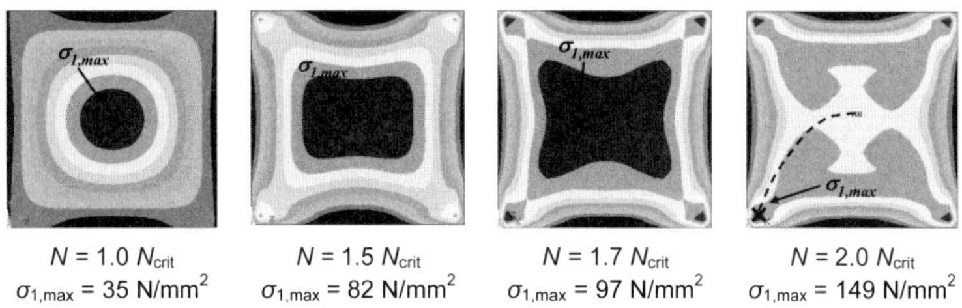

Figure 5.16: Typical maximum principal stress distribution on the surface of a buckled glass plate as a function of the load level.

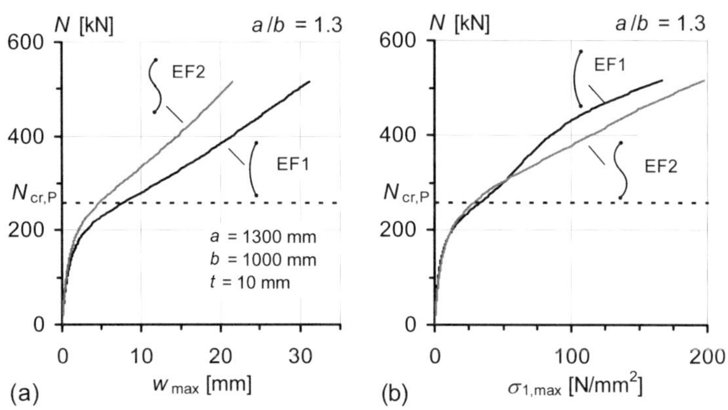

Figure 5.17: Influence of the buckling shape on the load N as a function of: (a) the maximum deflection w_{max}, (b) the maximum principal stress $\sigma_{1,max}$.

- on the front side, along the diagonal;
- on the reverse side, in the line of the maximum buckle bending; and
- in the anchor points of the diagonal tension field.

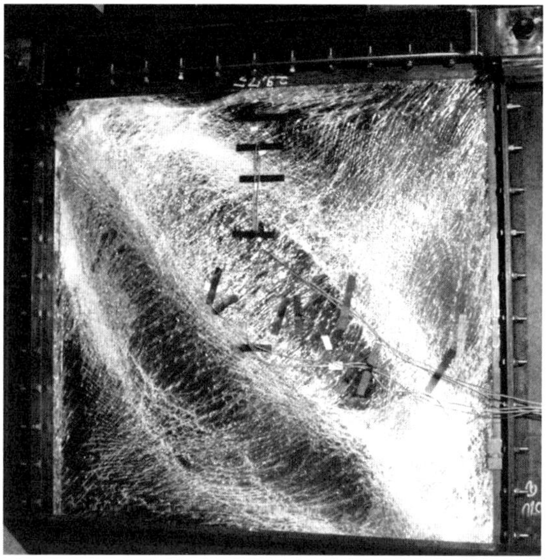

Figure 5.18:
Typical breakage pattern of a glass panel under in-plane shear load.

5.5.3 Structural design

In order to determine the buckling resistance of a glass panel, the distribution of the maximum principal stresses on the glass surface has to be known. Existing analytical plate buckling models are not precise enough to describe the stress distribution due to the non-linear behaviour. As long as design methods such as buckling curves do not exist for glass, finite element models are recommended for design.

Finite element plate buckling analysis is very laborious in practice; therefore the possibility of a design method based on buckling curves similar to the European steel design code [142] was investigated in [243, 338]. Typically, a buckling curve gives a reduction factor

$$\rho = \rho(\overline{\lambda}_{\mathrm{P}}) \tag{5.31}$$

which indicates the buckling resistance of the plate as a function of the slenderness ratio $\overline{\lambda}_{\mathrm{P}}$, which again characterizes the risk of the plate to buckle. The slenderness ratio for a plate subjected to in-plane compression is defined as

$$\overline{\lambda}_{\mathrm{P}} = \sqrt{\frac{\sigma_{\mathrm{Rk}} t}{N_{\mathrm{x,crit}}}} \tag{5.32}$$

and for in-plane shear loads as

$$\overline{\lambda}_{\mathrm{P}} = \sqrt{\frac{\tau_{\mathrm{Rk}}}{\tau_{\mathrm{crit}}}} \tag{5.33}$$

The characteristic buckling resistance of a glass panel may be defined as

$$N_{\mathrm{Rk}} = \rho \sigma_{\mathrm{Rk}} t b \tag{5.34}$$

for pure compression and

$$V_{\mathrm{Rk}} = \rho \tau_{\mathrm{Rk}} t b \tag{5.35}$$

for shear loads. The glass thickness is represented by t and b represents the width of the panel. The slenderness and the reduction factor are based on the characteristic values of the tensile strength σ_{Rk} and of the shear stress resistance τ_{Rk}. Wellershoff [338] proposes to simplistically assume $\tau_{\mathrm{Rk}} = \sigma_{\mathrm{Rk}}$.

The plate buckling verification is then performed with

$$N_{\mathrm{Ed}} \leq \frac{N_{\mathrm{Rk}}}{\gamma_{\mathrm{M}}} \quad \text{and} \quad V_{\mathrm{Ed}} \leq \frac{V_{\mathrm{Rk}}}{\gamma_{\mathrm{M}}} \tag{5.36}$$

where N_{Ed} and V_{Ed} are the design values of the applied force and γ_{M} is the partial safety factor for glass.

Reduction factors for different types of loading, glass geometries, initial deformations w_0 and boundary conditions were calculated by Luible [243] and Wellershoff [338] based on finite element models and plotted in buckling diagrams (e. g. *Figure 5.19*). These simulation results are a first step towards a future definition of plate buckling curves.

Figure 5.19: Simulated plate buckling reduction factors for monolithic and laminated glass subjected to a uniform pressure on the glass edge.

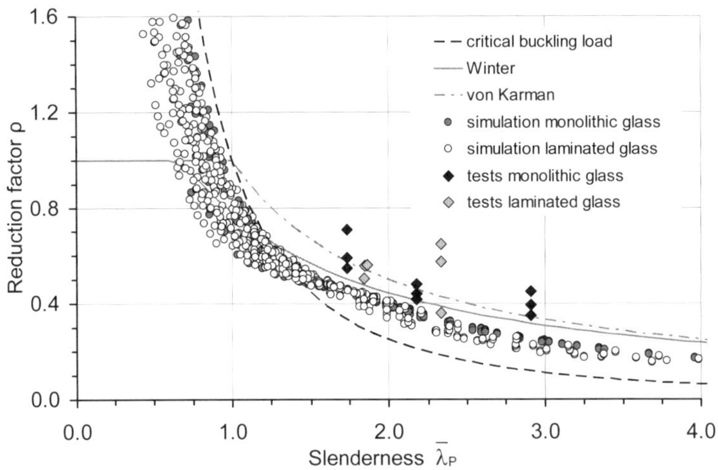

Reduction factors for pure compression

The simulations in [243] showed that it is possible to establish buckling curves for glass panels under pure compression. Furthermore the following conclusions can be drawn:

- The applied compressive edge stress σ_{p} can exceed the characteristic tensile glass strength σ_{Rk} because the compressive strength is much higher than σ_{Rk}. As the reduction factor is based on the tensile strength, ρ may become > 1. (The reduction factors used for steel construction are based on the yield strength and thus always < 1.)

- For a slenderness ratio $\overline{\lambda}_p > 1.5$ the design curves are almost independent of the initial deformation.
- For a slenderness ratio $\overline{\lambda}_p < 1.5$ the initial deformation w_0 has an influence on the buckling strength. The curves in this slenderness range have, therefore, to be defined as a function of the initial deformation.
- Plate buckling curves in steel construction are based on the assumption that the vertical edges are restrained in the plane and that horizontal edges are subjected to a uniform deformation. This cannot be assumed for glass panels. The vertical glass edges (e. g. in a glazing bead) are not restrained laterally in the plane and the load introduction (usually with soft materials) creates neither a uniform displacement nor a constant pressure on the glass edge. Therefore, the real behaviour lies between model (a) and model (b) in *Figure 5.15*. For this reason reduction factors for glass tend to be smaller than their equivalents for steel.
- Further research is required before a definitive definition of buckling curves for glass panels under pure compression is possible.

Reduction factors for shear loads

Reduction factors for glass panels under shear loads are studied in [338]. Based on the Ayrton-Perry-Format (similar to EN 1993-1 [142]), the following reduction function is proposed:

$$\rho = \begin{cases} 1 & \text{for} \quad \overline{\lambda}_p \leq \overline{\lambda}_0 \\ \left(\Phi + \sqrt{\Phi^2 - \overline{\lambda}}\right)^{-1} & \text{for} \quad \overline{\lambda}_p > \overline{\lambda}_0 \end{cases} \tag{5.37}$$

with

$$\Phi = \frac{1}{2}\left(1 + \alpha(\overline{\lambda} - \overline{\lambda}_0) + \overline{\lambda}\right) \tag{5.38}$$

Based on tests, it is recommended to use $\overline{\lambda}_0 = 0.8$ and $\alpha = 0.49$ for glass panes under in-plane shear forces. For combined shear loads and lateral loads, a design method is proposed in [338]. It is based on an interaction formula that was derived from finite element simulations and accounts for the shear and bending capacity of the panel.

Chapter

6 Design Methods for Improved Accuracy and Flexibility

6.1 Introduction

As mentioned in Section 4.6, many of the shortcomings of current standards, guidelines and design methods can be addressed with the generalized lifetime prediction model that was discussed in Section 3.3. This chapter provides an outline of this approach by summarizing the recommendations of Haldimann [189]. For more details, the reader can refer to this document.

6.2 Surface condition modelling

The lifetime prediction model described in Section 3.3 offers two alternatives for modelling a glass element's surface condition: a *single surface flaw (SSF)* and a *random surface flaw population (RSFP)*.

For structural design, it is essential to know which of these models to use and when. The characteristics and particularities of these two surface condition models are, therefore, discussed in the ensuing text. On this basis, recommendations for design and testing are given in Section 6.3.

6.2.1 Single surface flaw model

The surface condition of as-received glass can be characterized accurately by an RSFP, i. e. a large number of flaws of random depth, location and orientation (cf. Section 3.3.5). If, however, a glass element's surface contains a single flaw (or a few flaws) that is substantially deeper than the many small flaws of the RSFP, its resistance is likely to be governed by this deep flaw because it will initiate failure.

If the surface condition of a glass element can be represented by a SSF, its lifetime can be predicted by simulating the growth of this flaw using the equations derived in Section 3.3.4. A glass element is acceptable if a design flaw does not fail during the service life when the element is exposed to the design action history. In order to determine the stress history that the design flaw is exposed to, its location must be known. In some cases, it is possible to make a reasonable assumption about the location of the most severe flaws (e. g. in the vicinity of the bolt holes in bolted glass elements). In other cases, the location of the design flaw may be completely unknown. In such cases, it is safe to assume that the flaw is located anywhere on the surface, i. e. to simulate crack growth using the 'worst' stress history (the one that causes the most crack growth) that exists on the element's surface.

The SSF model caters to arbitrary geometries and loading conditions, as long as sensible assumptions with regard to the location of the design flaw and the crack opening stress history at this location can be made. Since the outcome of the model is a function of the conditions at the location of only the design flaw(s), it is not influenced by the element's size or by biaxial or non-homogeneous stress fields (in contrast to the RSFP model, see Section 6.2.2). Because of the simple representation of the surface condition, the model is intuitive, easy to use and numerical modelling is simple and fast. Furthermore, no statistical representation of the surface condition is integrated into the model. This is an advantage compared to RSFP-based modelling, because any statistical model (including discontinuous functions) considered appropriate for a specific task at hand can be used.

6.2.2 Random surface flaw population model

With this approach, the surface condition of a glass element is represented by a RSFP, i. e. by a large number of flaws whose number, location, orientation and depth are all represented by statistical distribution functions (cf. Section 3.3.5).

The lifetime of a glass element is predicted by simulating the growth of its surface flaw population under the influence of the design action history using the equations derived in Section 3.3.5. An acceptable design is achieved when the probability of failure during the service life is less than or equal to the target failure probability.

The model provides a good representation of as-received glass and glass with artificially induced homogeneous surface damage. It may, however, be unrepresentative of in-service conditions, especially if deep surface flaws are present on the glass surface or if glass elements contain machining damage. The model caters to arbitrary geometries and loading conditions, as long as the relevant crack opening stress history at all points on the surface can be determined. The approach accounts for the element's size and for biaxial and non-homogeneous stress fields. Surface damage hazard scenarios (see Section 2.1.2), however, cannot easily be modelled. Numerical modelling is generally complex and computing time intensive.

Random surface flaw population-based modelling yields accurate results for medium to high failure probabilities. The approach is, therefore, well suited to the interpretation of laboratory tests; however, it is less appropriate for structural design. The notable drawback is that extrapolation is required for the very low failure probabilities used in structural design, which is very sensitive to the scatter of the underlying strength data and to the choice of the target failure probability. This inevitably leads glass designers to adjust the target failure probability and the model parameters in order to obtain results close to experience values rather than to design on a proper physical basis.[1]

[1] This issue is well illustrated by European and North American standards. The two standard families adopt very different target failure probabilities. GFPM-based methods use a high target failure probability in

6.3 Recommendations for design

Structural design of glass elements with the lifetime prediction model generally involves the following steps (cf. *Figure 6.1*):

1. Decide whether a single surface flaw (SSF) or a random surface flaw population (RSFP) is the more suitable surface condition model for the task at hand.
2. Determine the design crack depth $a_{i,d}$ (SSF) or the surface condition parameters θ_0 and m_0 (RSFP) by testing at inert conditions (see Section 6.4) or using another suitable method (engineering judgement, flaw detectability criteria, etc.).
3. Make conservative assumptions for the crack velocity parameters (n, v_0), the fracture mechanics parameters (K_{Ic}, Y) and the residual surface compression stress (σ_r).
4. Define a design action history and establish the action/stress relationship (normally by finite element analysis) for the location of the design flaw (SSF) or for all points on the element's surface (RSFP).
5. Assess the structural performance of the glass element and modify the design if required.

The following paragraphs explain how the above-listed design steps may be applied to glass elements encountered in practice. It will be seen that only a few cases actually require all of the above-mentioned steps to be carried out.

Exposed glass surfaces

- **Definition.** Exposed surfaces are glass surfaces that may be exposed to accidental impact, vandalism, heavy wind-borne debris or other factors that result in surface flaws that are substantially deeper than the 'natural' flaws caused by production and handling. Such flaws will be called 'severe damage' hereafter.
- **Surface condition model.** Structural design of glass elements with exposed surfaces should be based on a *design flaw*, which is a realistic estimation of the potential damage caused by surface damage hazards. Accordingly, the surface condition should be represented by a single surface flaw (cf. Section 6.2.1).
- **Long-term loading**. Long-term inherent strength[2] in the presence of a deep design flaw is generally low (see *Figure 3.6*), has a large scatter and depends on many external influences (see Section 3.2). Therefore:

combination with ambient strength data from weathered window glass specimens. European methods use the low target failure probability required by EN 1990:2002 [135]. Therefore, the latter cannot use strength data from weathered window glass because this would yield an unrealistically low design resistance. To avoid this problem, the European methods employ ambient strength data from specimens with artificially induced homogeneous surface damage. Compared to the damage on weathered window glass, the homogeneous damage reduces the scatter of strength data markedly. As a consequence, the surface condition parameter m_0 becomes high enough to allow for the use of a low target failure probability without obtaining unrealistic results. In real terms, these adjustments yield very similar results at low probabilities of failure as the design initial crack depths that the standards implicitly assume to be present on the surface of a glass element are in fact very similar.

For a detailed discussion of this issue, see [189, Section 8.3.2].

[2]Definition, see Section 3.3.2.

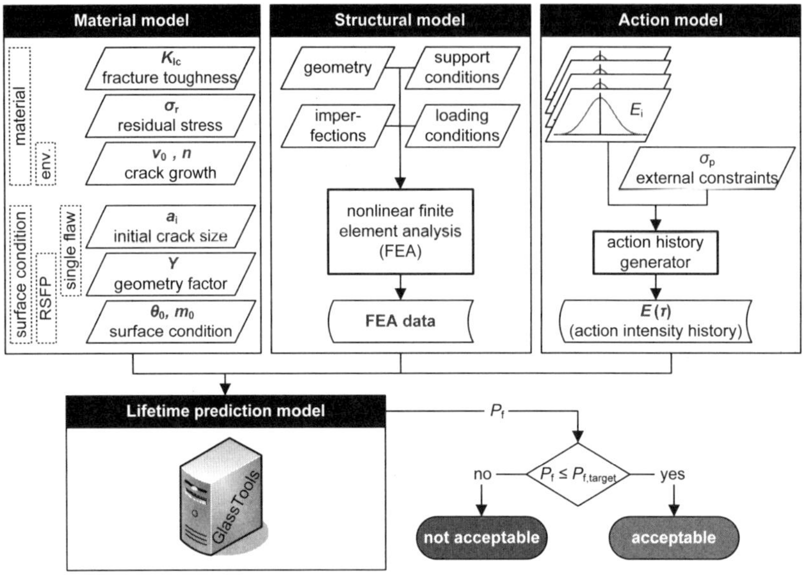

Figure 6.1: Structural design of glass elements with the lifetime prediction model [189].

- ◇ *Annealed glass* should not be relied upon for structural glass elements with long-term tensile loads and exposed surfaces. If annealed glass must be used for some reason (cost, optical quality, tolerances, element size, etc.), failure consequences have to be evaluated very carefully. Protection of building occupants in the case of glass breakage, post-breakage structural capacity, structural redundancy and easy accessibility for the replacement of broken glass elements become key aspects.
- ◇ In the case of *heat strengthened or fully tempered glass*, the inherent strength in the presence of a design flaw is low compared to the residual stress, so that it provides only a minor contribution to the effective resistance. In view of this limited structural benefit and the complex time-dependent behaviour, it is reasonable to ignore the inherent strength entirely and to design the glass element such that surface decompression is prevented at all points on the surface and during the entire service life (cf. Section 3.3.2). Because the residual stresses are independent of service-life conditions such as stress history, environmental conditions etc., this kind of design is extremely simple.

- **Impact and short-term loading.** While neglecting the inherent strength when designing heat strengthened or fully tempered glass is safe, it may in some cases be deemed too conservative for impact and short-term loads. In these cases, the inherent strength can be estimated as described above for annealed glass.
- **Quality control and inspections.** In critical applications, it may be possible to use information from inspections for periodically assessing the strength of the glass elements. This information may be obtained by undertaking periodic inspections during the entire service life. In the case of heat-treated glass, it is more effective and economical to improve the quality control measures during the tempering process and ignore the inherent glass strength (with or without inspection).

Non-exposed glass surfaces

- **Definition.** *Non-exposed surfaces* are glass surfaces that are permanently and safely protected in all relevant hazard scenarios, so that they do not undergo any surface damage from external influences. Examples are inwardly oriented faces of laminated glass and insulating glass units, the surfaces of inner sheets of triple-laminated glass and surfaces of glass elements that are installed in sheltered locations.
- **Surface condition model.** RSFP-based modelling (cf. Section 6.2.2) is suitable for glass elements with non-exposed surfaces, as long as loading conditions are rather simple and failure away from the edges is relevant. If edge failure is relevant or if complex loading conditions make the RSFP-based model too complex to be used with reasonable effort, SSF-based modelling is a conservative and much simpler alternative.
- **Quality control and inspections.** Less conservative design is possible if it is based on a relatively low maximum surface flaw depth, which is ensured by inspection, and if required replacement, of the structural elements immediately after installation. Although the inherent strength is much higher compared to that of exposed elements, improving quality control measures during the tempering process is likely to be a more economical means of improving the design strength of the glass elements.

Machining damage

Since the orientation and, more importantly, the location of the flaws are often not of a random nature, an RSFP-based model could produce unsafe results if glass elements contain significant machining damage. It is, therefore, recommended to design such elements using a design flaw that accounts for both machining damage and surface damage hazard scenarios.

Non-structural glass elements

If failure and replacement of an element in the case of severe surface damage are accepted, non-structural elements can be designed as non-exposed structural elements. If non-structural elements have to withstand mainly lateral loads, design is, especially for heat-strengthened or fully tempered glass, often governed by deflection criteria.

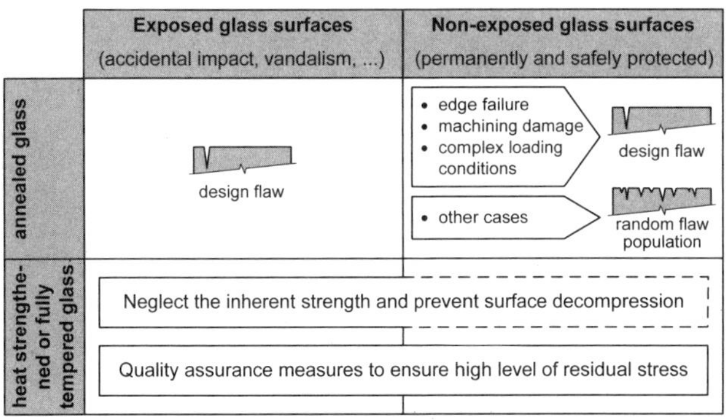

Figure 6.2: Summary of the recommendations for structural design given in this chapter.

6.4 Testing

6.4.1 Introduction

Testing is required mainly for two reasons:

1. To determine parameters of predictive models and design methods.
2. To verify or augment the predictive calculation. This is typically related to cases where structural glass design cannot be solely based on predictive modelling. The difficulties with modelling arise mainly in the following areas:

   a) Glass is extremely sensitive to *stress concentrations*. Numerical models, however, often cannot provide reliable information on stress fields and particularly stress concentrations. This lack of confidence in the numerical models often arises when there is limited information about the materials being modelled (e. g. liners, gaskets, bushings) and / or when the assembly process may cause stress raising imperfections (e. g. misalignment, large tolerances).

   b) Despite recent advances in the field [231], the *post-breakage structural capacity* often cannot be reliably predicted by predictive modelling.

   c) There is not much experience and quantitative information available concerning the *surface damage* caused by various hazard scenarios.

   d) The response of structural elements or entire sub-structures to *impact loads* is difficult to model.

   e) Building owners, insurers and authorities generally have *little confidence* in glass structures and often ask for full scale tests.

In particular the following issues should be considered:

- It is very important that design and interpretation of tests are based on a thorough understanding of the material behaviour. The fact that results from tests at ambient conditions represent a combination of both surface condition and time-dependent crack growth is particularly crucial. It is unfortunate that much project-specific testing is performed without taking time-dependent effects properly into account.
- If testing at ambient conditions is unavoidable, sub-critical crack growth during the tests must be modelled. While this can efficiently be done using the model from Section 3.3, dependence of the crack velocity parameters on the environmental conditions and the stress rate further diminishes the accuracy and reliability of the results.
- The problems related to sub-critical crack growth in laboratory tests can be addressed by the near-inert testing procedure summarized in Section 6.4.2. By preventing sub-critical crack growth during tests, it allows substantial improvement in the accuracy and safety of test results.
- Tests on as-received specimens or on specimens with artificially induced homogeneous surface damage are unsuitable for assessing the structural performance of glass elements in surface damage hazard scenarios. Such elements should be tested with realistic design flaws. This issue is discussed in Section 6.4.3.

6.4.2 Determination of surface condition parameters

Introduction

Reliable surface condition parameters (θ_0, m_0) form the basis of RSFP-based modelling and must be derived from glass strength data. The testing procedures used today to obtain glass strength data were explained in Section 3.5.2. While European and North American design methods are based on fundamentally different testing procedures (see also *Table 4.8*), all current design methods use strength data obtained at *ambient conditions*, i.e. in normal, humidity containing air. This strength data depends on a specimen's surface condition *and* on the sub-critical growth of the surface flaws during the tests. It was shown in Section 3.2.3 that the relationship between stress intensity and crack velocity varies widely and depends strongly on the environmental conditions, on the residual stress in the glass and on the stress rate at which a specimen is loaded. This prevents accurate estimation of the growth of surface flaws during experiments. Inaccurate estimation, however, can result in unsafe design parameters. *Glass surface condition data should, therefore, be obtained from laboratory testing in near-inert conditions.*

Creating near-inert conditions in laboratory tests

Considering the chemical background of stress corrosion (see Section 3.2), inert conditions can be achieved in various ways:

1. Testing in a vacuum or in a completely dry environment.
2. Testing in a normal environment with a hermetic coating.
3. Testing in a normal environment at very rapid stress rates.
4. Testing at a sufficiently low temperature, at which the kinetics of environmentally induced reactions are arrested.

Not all possibilities outlined above are equally suitable for structural applications. Options 1 and 4 are difficult and expensive, especially for full-scale testing on large specimens. Options 2 and 3, in contrast, are comparatively simple and inexpensive provided that the conditions do not need to be fulfilled perfectly. Haldimann [189] showed that near-inert conditions in laboratory tests can be achieved by combining a near-hermetic surface coating and a relatively rapid stress rate. The latter reduces the effect of the former's imperfection and vice versa. The proposed testing procedure is as follows:

1. **Drying**. The specimens are dried in an oven at $100 \pm 5\,°C$ for 48 ± 6 hours. The humidity in the oven is maintained below 5% RH by a high-performance molecular sieve desiccant.
2. **Hermetic coating**. To achieve a hermetic coating, a silicone grease is applied to the tension face of the specimens. This grease is highly hydrophobic, impermeable and its viscosity is high enough to ensure that the coating remains intact during handling and testing.
3. **Adoption**. Specimens are kept at ambient conditions for 2 hours to allow them to adopt ambient temperature.
4. **Destructive testing**. The specimens are loaded to failure using a high stress rate (about 20 MPa/s is recommended).

If some sub-critical crack growth occurs during near-inert tests, the results are conservative. This is a major advantage over ambient testing, in which overestimation of the crack growth during the tests leads to too optimistic surface condition parameters and therefore to unsafe design.

Interpretation of experimental inert strength data

The experimentally determined failure stresses at inert conditions represent the material's inert strength. The surface condition parameters can be obtained from such data as follows:

- For *tests with simple stress fields*, such as coaxial double ring tests or four-point bending tests, simple analytical equations can be used (see [189, Section 5.3.3]). Fitting of the Weibull distribution to test results can be done by simple parameter estimation or maximum likelihood fitting.
- For *tests with complex stress fields*, such as tests on large rectangular glass plates, the general lifetime prediction model described in Section 3.3 should be used. Failure loads and stresses are influenced by the non-linear load/stress relationship and the location-dependent stress history caused by the complex stress field. Experimental data, which at best provides information about the stress history at a few discrete points on the surface, will therefore generally not follow a Weibull distribution, such that a distribution-independent fitting method such as least-squares fitting or maximum likelihood fitting should be used to determine surface condition parameters.

  An example of how this can be done as well as the required algorithms and their implementation in computer software are provided in [189].

Using strength data from ambient tests

If no inert strength data is available, the derivation of surface condition parameters from ambient strength data may be useful. Equations required for this purpose are given in [189]. They are derived from Equation (3.42) by narrowing its range of validity to constant stress or constant and moderate stress rate, uniform stress fields, a constant principal stress ratio and constant crack velocity parameters.

6.4.3 Obtaining strength data for design flaws

Current design methods rely on strength data obtained from specimens with as-received surfaces, specimens with artificially induced homogeneous surface damage or weathered

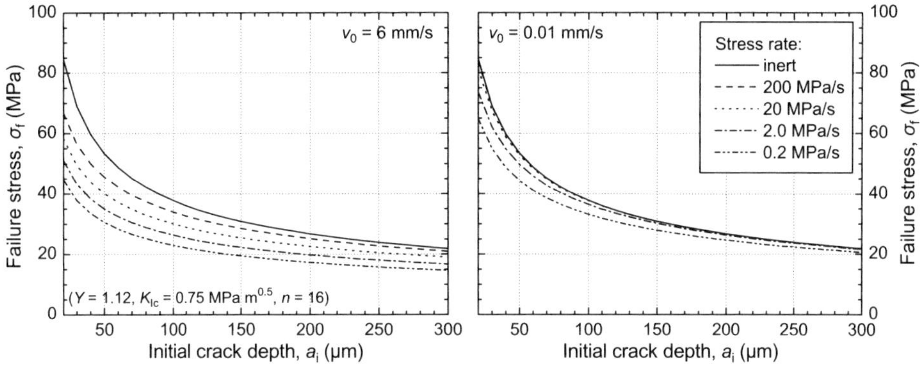

Figure 6.3: Failure stress of surface flaws in constant stress rate tests. In common laboratory conditions (right graph), the strength of deep surface flaws measured at ambient conditions and with a stress rate of 20 MPa/s or above is virtually identical to the inert strength. This was confirmed by experiments in [189].

specimens. Specimens with such surface conditions are useful when adopting a RSFP-based approach (see Section 6.2 and 6.3). Such strength data is, however, unsuited for design flaw-based design (see the same two sections). To obtain strength data for this approach, i. e. to quantify the damage caused by a surface damage hazard scenario (design flaw) and to assess the structural performance of glass elements that contain such damage, tests need to be performed on specimens with deep close-to-reality flaws. Such flaws have to meet two conflicting requirements:

1. They should be as similar as possible to the surface damage that structural glass elements are likely to undergo under in-service conditions. This includes accidental damage (e. g. due to handling, cleaning, impact of vehicles, tools falling down or impact of heavy wind-borne debris) as well as malicious damage (vandalism).
2. They should be as reproducible as possible.

In order to achieve an optimal compromise between these requirements, Haldimann [189] suggests to use a specially developed surface scratching device. This device may be used to induce long surface cracks on the glass surface by applying a constant force to a 0.33 carat dressing diamond. This scratching tip was found to be well suited because it does not show much wear and because its relatively large opening angle causes some widening of the scratch, which is an effect that is likely to happen with objects commonly used by vandals (e. g. diamond rings). A steel plunger holds the diamond tip. A casing guide ensures that the plunger is positioned exactly perpendicular to the glass plate. Ball bearings are used to minimize the sliding friction between the plunger and the guide.

The plunger can be loaded with steel blocks of known weight and creates a constant contact pressure between the scratching tip and the specimen. In dry diamond on glass scratching, the regularity of the surface flaws is problematic. An evaporating glass cutting oil makes depth and geometry of the flaws more uniform and allows higher loads to be applied.

The following should be considered with regard to deep surface flaw testing (for more details, see [189, 190]):

- The scatter of the strength of deep surface flaws is extremely high.
- The locally fractured glass zone around a surface scratch is significantly deeper than the open, visible depth. The effective nominal flaw depth that governs strength is, therefore, significantly deeper than the optically measured flaw depth. This phenomenon is less pronounced in heat-treated glass where the residual compressive stresses hinder fracture of the glass beyond the zone that is in direct contact with the scratching tip. Therefore, the strength reduction caused by a given surface damaging influence is much less severe in heat-treated glass than it is in annealed glass.
- When testing specimens with deep surface flaws, the time to failure is so short that a high stress rate is sufficient to ensure near-inert conditions (see *Figure 6.3*). Strength measurements obtained this way, i. e. without drying and hermetic surface coating, can be interpreted as inert strength data without being excessively conservative. This makes such laboratory testing simple and inexpensive, even in the case of large structural elements.[3]
- The key factor for meaningful results is a close match between the design flaw and potential in-service damage.

6.5 Overview of mathematical relationships

An overview of the mathematical relationships to be used in the different cases is provided in *Table 6.1*.

[3] Because of the longer time to failure and the small crack depth, the same is not true for as-received glass specimens.

Table 6.1: Table of mathematical relationships (SSF = single surface flaw, RSFP = random surface flaw population).

Structural design (simplified design equations)			
SSF	$\sigma_{t_0} \leq \sigma_{\mathrm{R},t_0}$	$\sigma_{\mathrm{R},t_0} = \left(\frac{1}{t_0} \frac{2}{(n-2) \cdot v_0 \cdot K_{\mathrm{Ic}}^{-n} \cdot (Y\sqrt{\pi})^n \cdot a_{\mathrm{i}}^{(n-2)/2}} \right)^{1/n}$	→ Equation (3.23)
		$\sigma_{t_0} = \left(\frac{1}{t_0} \int_0^T \sigma^n(\tau)\,\mathrm{d}\tau \right)^{1/n} \approx \left(\frac{1}{t_0} \sum_{j=1}^{J} \left[\sigma_{t_j}^n \cdot t_j \right] \right)^{1/n}$	→ Equation (3.22)
RSFP	$\bar{\sigma} < f_0$	$f_0 = \left[-\ln(1 - P_{\mathrm{f,t}}) \right]^{1/\bar{m}} \cdot \left(\frac{t_0}{U \cdot \theta_0^{n-2}} \right)^{-1/n} = f_{0,\mathrm{inert}}^{(n-2)/n} \cdot \left(\frac{U}{t_0} \right)^{1/n}$	→ Equation (3.50)
		General cases:	
		$\bar{\sigma} = \left(\frac{1}{A_0} \int_A \sigma_{1,t_0}^{\bar{m}}\,\mathrm{d}A \right)^{1/\bar{m}} \approx \left(\frac{1}{A_0} \sum_i \left[A_i \cdot \sigma_{1,t_0,i}^{\bar{m}} \right] \right)^{1/\bar{m}}$	→ Equation (3.46)
		$\sigma_{1,t_0,i} = \left(\frac{1}{t_0} \int_0^T \sigma_{1,i}^n\,\mathrm{d}\tau \right)^{1/n} \approx \left(\frac{1}{t_0} \sum_{j=1}^{J} \left[\sigma_{1,\tau_j,i}^n \cdot \tau_j \right] \right)^{1/n}$	→ Equation (3.43)
		Simplification if the decompressed surface area remains constant and the major principal stress is proportional to the load at all points on the surface and during the entire loading history:	
		$\bar{\sigma} = A_0^{-1/\bar{m}} \cdot \mathbb{X}_{t_0} \cdot \bar{A}^{1/\bar{m}}$ with $\bar{A} = \int_A c(\vec{r})^{\bar{m}}\,\mathrm{d}A \approx \sum_i A_i c_i^{\bar{m}}$	→ Equation (3.53)
		$\mathbb{X}_{t_0} = \left(\frac{1}{t_0} \int_0^T \mathbb{X}^n(\tau)\,\mathrm{d}\tau \right)^{1/n} \approx \left(\frac{1}{t_0} \sum_{j=1}^{J} \left[\mathbb{X}_{t_j}^n \cdot t_j \right] \right)^{1/n}$ with $\mathbb{X} = \begin{cases} \breve{\sigma} & \text{for stresses} \\ q & \text{for loads} \end{cases}$	→ Equation (3.57)
Interpretation of tests (general equations for predictive modelling)			
SSF	$\tilde{a}_c$	$\tilde{a}_c(\tau) = \left[\left(\frac{\sigma_{\mathrm{n}}(\tau) \cdot Y\sqrt{\pi}}{K_{\mathrm{Ic}}} \right)^{n-2} + \frac{n-2}{2} \cdot v_0 \cdot K_{\mathrm{Ic}}^{-n} \cdot (Y\sqrt{\pi})^n \cdot \int_0^\tau \sigma_{\mathrm{n}}^n(\tilde{\tau})\,\mathrm{d}\tilde{\tau} \right]^{\frac{2}{2-n}}$	→ Equation (3.21)
RSFP	P_{f}	General cases:	
		$P_{\mathrm{f}}(t) = 1 - \exp\left\{ -\frac{1}{A_0} \int_A \frac{2}{\pi} \int_{\varphi=0}^{\pi/2} \left[\max_{\tau\in[0,t]} \left\{ \left(\left(\frac{\sigma_{\mathrm{n}}(\tau,\vec{r},\varphi)}{\theta_0} \right)^{n-2} + \frac{1}{U \cdot \theta_0^{n-2}} \int_0^\tau \sigma_{\mathrm{n}}^n(\tilde{\tau},\vec{r},\varphi)\,\mathrm{d}\tilde{\tau} \right)^{\frac{1}{n-2}} \right\} \right]^{m_0} \mathrm{d}A\,\mathrm{d}\varphi \right\}$	→ Equation (3.42)
		For simplifications for common test setups with uniform stress fields and for testing at inert conditions, see [189, Section 5.3.3].	

Chapter

7 Glass Connections

7.1 Introduction

The traditional approach for dealing with connections between glass and other materials was to avoid direct contact between the glass and other harder materials thereby diverting loads or movement away from the glass. Although this sound engineering advice still holds true today, the past 25 years has seen an increasing architectural trend to maximize transparency when using glass. This trend can be traced through the chronological development of glass connections: from the linearly supported glazing associated with the curtain walls developed in the mid 20th century, to the patch plate friction fittings developed in the mid-1970s, to the bolted point supports developed in the 1980s and 1990s (*Figure 7.1*).

These developments show a gradual reduction in the size of the glass support and an increase in the magnitude and types of loads that are transmitted to the glass. In all

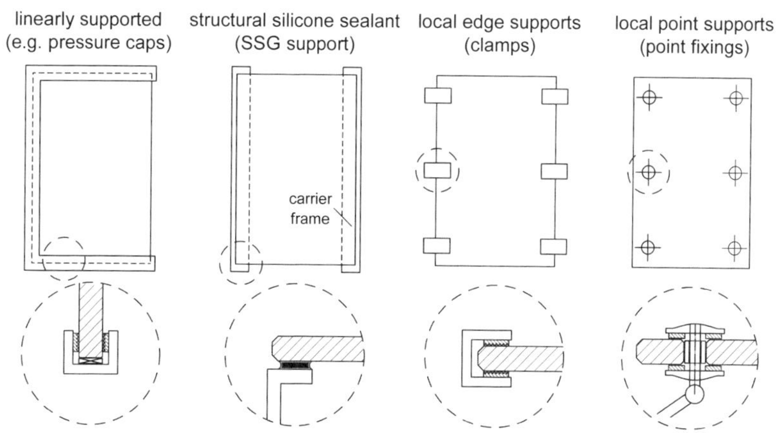

Figure 7.1: Summary of common glass support types.

these connections the direct contact (or hard spots) between glass and harder materials should still be avoided by employing intermediate materials. These intermediate materials often have a smaller or comparable stiffness to glass, but should have the necessary material strength and stiffness to transfer the loads and also have an adequate durability. Suitable intermediate materials are plastic, resins, neoprene, injection mortars, aluminum or fibrous gaskets.

More recently there have been promising developments in chemical or glued connections in glass. This has opened up a range of exciting possibilities that was not possible with mechanical connections, but at the same time a series of associated problems such as durability of the adhesive joints must now be considered.

These developments in glass connections mean that the engineer is now faced with a wide range of possible techniques and products for connecting glass-to-glass or glass to other materials. The aim of this chapter is to provide a general overview of these techniques and to provide guidelines on their correct application.

For the purposes of this document it is convenient to distinguish between two main types of connections, namely, mechanical connections and adhesive connections. In some cases the connection may be a combination of a mechanical and an adhesive connection. Such combined connections may improve the performance of the joint, however, in cases where stiff adhesives are employed, the adhesive element of the joint is often substantially stiffer than the mechanical part of the joint. Consequently the adhesive will carry the majority of the loads and the mechanical connection will only come into effect once the capacity of the adhesive has been exceeded.

7.2 Mechanical fixings

This text has been compiled in collaboration with the following experts:
Christoph HAAS, Benjamin BEER

7.2.1 Linearly supported glazing

Linearly supported glazing is often used in framed constructions, such as curtain wall systems, where rectangular glass panels are supported along two or four edges. The self-weight of the glass is transferred to the frame through plastic setting blocks located at the horizontal bottom glass edge. Alternatively the self-weight may be transmitted through neoprene layers with a Shore A hardness ranging between 60 and 80.

Lateral loads, normally arising from wind pressure and suction may be resisted mechanically by clamping the glass between the frame system on one side and a glazing bead or a capping/pressure plate on the other side (*Figure 7.2*). The loads are transferred from the glass to the framing system through 6 to 15 mm neoprene, EPDM or silicone gaskets (*Figure 7.2*). These supports allow a good degree of rotation of the glass edge and may consequently be considered as simple supports for the purposes of analytical and numerical modelling.

In framed systems, the frame size is larger than the glass pane. This clearance should be sufficiently large to accommodate both the induced deviations that result from manufacturing or construction tolerances and the inherent deviations that result from post-installation dimensional changes.

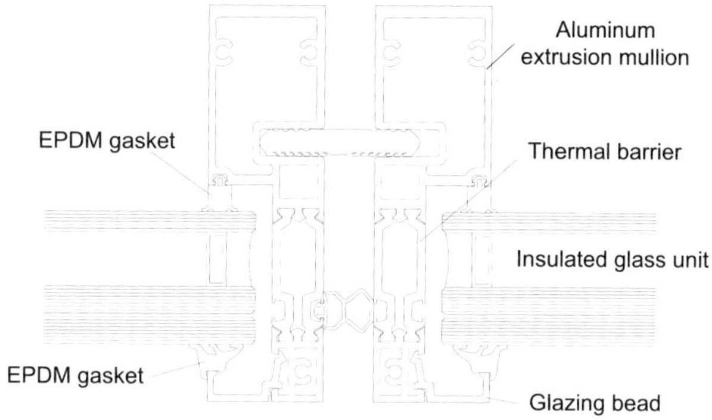

Figure 7.2:
Typical linear glass support with EPDM gaskets and glazing beads.

An alternative connection to mechanical fixing, also used in framed systems, is structural silicone glazing, which involves gluing the glass onto the frame system. This is discussed in further detail in Section 7.3.2.

Although less common, linear glass edge supports may also be used to transmit in-plane loads into the glass. This requires a higher degree of care and engineering at both design and construction stages and generally involves a bespoke system. In these cases the following recommendations should be considered:

- The areas close to the corners of heat-treated glass suffer from lower residual stresses. In order to prevent the glass corners from breaking, the load should therefore be introduced at a certain distance from the weaker corners (at least 2 times the glass thickness for heat strengthened glass (HSG) and fully tempered glass (FTG) and at least 100 mm for annealed glass (ANG)).
- The glass edges should be chamfered and ground or polished to avoid local stress concentration and premature failure of the glass.
- The compressive stress distribution on the glass edge is not constant but depends on the modulus of elasticity of the intermediate material and the sub-structure.
- Thermal movement of the sub-structure and the glass pane may create additional tensile stresses on the glass edge.
- The edge of laminated safety glass made of heat strengthened or tempered glass is normally not perfectly flush. The glass layers are tempered before laminating and mechanical edge treatment of tempered glasses is not possible afterwards. This leads to an asymmetric load introduction into the glass edge. In case of setting blocks, only one single glass layer would be supported while the other is not in contact with the setting block. A better solution is to employ special setting blocks with pin compensators or steel shoes, where the space between the steel shoe and the glass edge is filled with a special injection mortar (i. e. epoxy resin) or high strength glue to adjust the glass edges. In such a system all glass layers are supported. Nevertheless, variations in the thickness of the filling material may cause additional load eccentricities that have to be taken into account for the design of the laminated glass unit [243].

7.2.2 Clamped and friction-grip fixings

Clamped fittings were developed in order to minimize the visual impact of linear supporting frames and pressure cap profiles. Panel edges are fixed to the sub-structure

Figure 7.3:
Typical glass edge of annealed and heat strengthened or tempered laminated glass with possible load introduction.

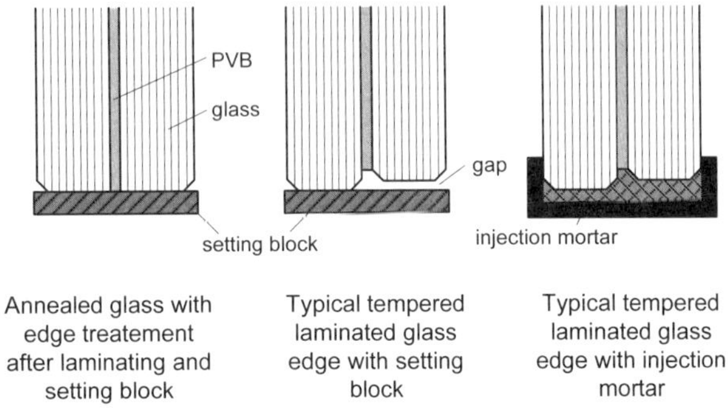

Figure 7.4:
Typical low-friction clamp fixing.

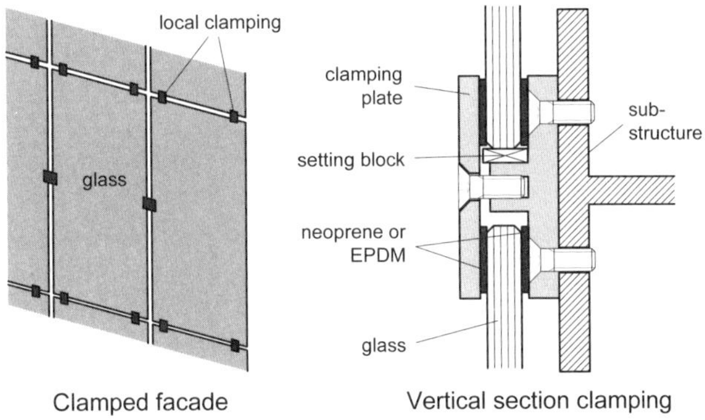

at discrete locations by means of clamps (*Figure 7.4*) which may be fixed back to an interior sub-frame or possibly to glass fins. Clamped fixings also facilitate the draining of water from overhead glazing where obstructions above the glass should be kept to minimum. Low friction clamped fixings are mainly used to transfer loads perpendicular to the glass pane. Setting blocks on the bottom glass edge allows for dead load support. In such clamped fittings the metal clamping plate simply holds the glass in place and is separated from the glass by a soft intermediate material such as neoprene or EPDM.

Other clamped fixings, however, are also able to transfer in-plane loads by clamping the fixings tightly to create a friction-grip connection. Friction-grip connections are theoretically well suited for the introduction of in-plane tensile loads because they distribute the load over a larger surface area than, say, a bolt-only arrangement and thus avoid major stress concentrations. Typically, the set-up consists of the glass pane, steel plates on both sides, gaskets between the steel plates and the glass and bolts which clamp the steel plates together. Direct contact between the glass and the steel parts are avoided by having oversized bolt holes and by employing the gaskets that act as an interlayer material between the glass and the steel plates (*Figure 7.5*). The gasket must be strong enough to withstand the normal stresses induced by the pre-stressed bolts without oozing out of the joint and must also resist the shear stresses induced by the in-plane force. At the same time it must not be too hard such as to damage the glass and it must also be sufficiently flexible to allow for fabrication tolerances between the glass and the steel plates. Additionally it should exhibit very low creep to prevent normal forces in bolts from decreasing over time. Typical gasket materials are pure aluminium or fibre gasket and are in the order of 1 mm thickness.

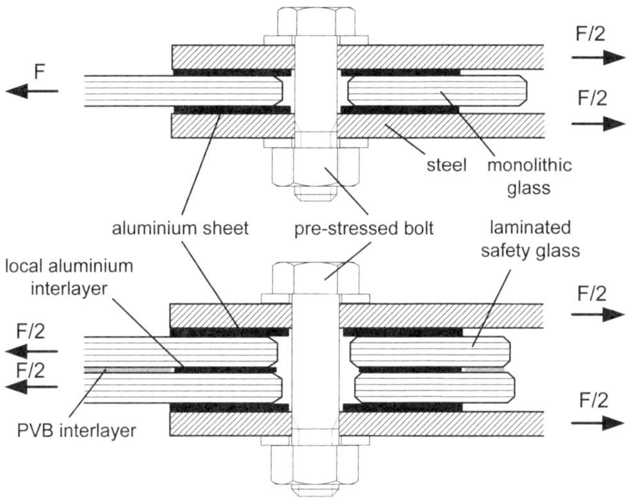

Figure 7.5: Typical friction grip connections: monolithic and laminated glass.

Special care should be taken when designing friction-grip connections in laminated safety glass. Interlayer materials in laminated safety glass such as polyvinyl butyral (PVB) are unable to withstand the clamping forces induced by the connection without oozing out and they suffer from large creep deformations that reduce the pre-stress over time. Consequently the interlayer in the region of the clamped connection is often removed and replaced by a stiffer, non-viscous material (e. g. aluminium alloys close to pure aluminium, which have a low modulus of elasticity) with the same thickness (*Figure 7.5*).

The force that can be transferred by friction depends on the geometry of the connection, the stiffness of the materials involved, the lowest coefficient of friction between the various interfaces and the long-term load-bearing capacity of the various components.

Panait [271] carried out experimental and theoretical studies on friction-grip bolted connections with monolithic glass and aluminum interlayer plates. In a first step the dry friction between glass and aluminum was investigated experimentally. The resulting friction grip resistance turned out to be time and temperature dependent. The coefficient of friction increases with time and with increasing temperature. These experimental findings were used to construct and validate numerical models of friction-grip connections.

Although research is ongoing in the field of friction grip connections, several applications for fins in facades [253] or beams (e. g. Glasgow Medical School) have been constructed with this type of connection. In general, these projects employ a combination of rules of thumb, finite element analysis and project-specific testing in order to determine the load-bearing capacity of these connections. Further information on this approach is provided in Section 8.1.

It is important to note that, depending on the glass geometry and clamp location, clamps may cause local rotational restraints in the glass which in turn result in stress concentrations at these locations. Unless a free rotation of the glass edge in the clamp fixing can be achieved in practise (i. e. by adopting a sufficiently thick and soft intermediate material) the restraint from the clamp must be considered in the analysis model.

7.2.3 Bolted supports

The use of discrete bolted connections is clearly not the most efficient way to transfer loads through a notoriously brittle material such as glass. This type of connection is often driven by aesthetic requirements to minimize the visual impact of the glass panel supports. Over the past 20 years, there have been various developments and refinements of bolted connections for glass structures. This has resulted in a wide variety of bolted connections. However, there are many similarities between the various bolted connections available. The first part of this section therefore provides general information and recommendations that apply across the board, this is followed by a description of some specific types of bolted connections.

If the members joined by a bolted connection consist of elasto-plastic materials (e. g. steel) the connection is generally able to redistribute the high bearing stresses around the bolt holes by yielding locally and thus exhibiting a high redundancy and a high load-bearing capacity. If this is extended to multiple bolts, the elasto-plastic materials also ensure a uniform load distribution on all bolts in the connection. In the case of brittle materials such as glass, the material is unable to redistribute local stress concentrations by yielding, consequently the high local stress concentrations in the bolt hole constitute a major problem.

Therefore one of the key challenges of structural detailing in glass is to devise a connection in which the high stress concentrations and direct steel-to-glass contact are avoided. This is in part achieved by intermediate materials in the form of bushings or liners that have a lower modulus of elasticity than glass. The materials used for these bushings should therefore be sufficiently strong and stiff to transfer loads to and from the glass without breaking or oozing out of the joint, but at the same time they should be soft enough to redistribute stress concentrations. An adequate resistance to creep and cyclic loading as well as a good UV-resistance is also important. Materials commonly used for bushings are aluminum, plastics such as EPDM (ethylene propylene diene monomer), PEEK (polyether ether ketone), POM (polyoximethylen) or polyamide or injected resin or mortar (e. g. HILTI HIT [195]). In this way the intermediate material is able to redistribute the compressive stress concentrations before they reach the glass. However, it is important to note that although such an approach has a major affect on the compressive stresses at the location of contact, it only has a minor affect on the tensile stresses caused by the elongation of the hole.

General performance and recommendations for bolted connections

The engineer must endeavour to reduce the stress concentrations by design whenever possible. Overend [263] and Maniatis [248] investigated the influence of several parameters on the structural behaviour of bolted connections with different bushing materials and for monolithic glass:

- The *closeness of fit* i. e. the bolt diameter relative to the hole diameter is directly related to the major principal stresses around the bolt hole. A larger clearance leads to higher maximum stresses in the glass hole and causes a shift in the position of the maximum stress. Compared to a tight-fit connection a clearance of 2 mm leads to an increase of the maximum principal tensile stress in the glass of about 66% for aluminium and 39% for POM-C bushings.

- The *geometry of the glass panel* particularly the glass thickness and the edge and end distances from the bolt hole to the glass perimeter have a major influence on the stress distribution around the glass hole. A thinner glass panel and small edge or end distances reduce the cross-sectional area of glass available to resist the load and thereby resulting in higher stresses.
- The *bushing material* has an influence on the magnitude of the maximum principal tensile stress around the glass hole, however this influence is reduced to a negligible level for tight-fitting connections.
- The *friction between bushing material and glass* has an influence on the maximum principal stress.
- An *eccentric load application* may also significantly increase the maximum principal stress in the glass hole.

The quality of the glass surface and the residual stress in the hole (cf. Section 3.6) also have a major influence on the load bearing capacity of bolted glass panels. Since the maximum tensile stress often occurs close to the holes, a realistic analysis model and careful detailing are essential for the design of bolted glass.

It is rarely possible to determine the stress distribution around the bolt holes by using simple formulae or charts. However, when the bolted connection is subjected to simple horizontal shear (e. g. when the splice is subjected to direct tension or compression such as with through bolt connections, see Section 7.2.3) the maximum principal stresses may be determined by applying stress concentration factors [275].

However, these stress concentration factors simply provide a single value for the maximum principal stress and do not provide information such as the stress distribution around the hole or at a distance from the hole edge, which is essential for determining the load-bearing capacity of the glass connection in an accurate manner. A 3-dimensional finite element model may be used to provide the full stress distribution around and away from the bolt hole. Guidance on the use of finite element method is beyond the scope of this document and readers should refer to the numerous publications on the subject, such as Zienkiewicz and Taylor [352] and Cook [73]. However, Siebert [315] and Overend and Maniatis [267] give some guidelines for good modelling, which are summarized here:

- The entire bolted assembly including the intermediate material and their respective mechanical properties should be modelled.
- Gap or contact elements should be used around the bolt hole to ensure that only compressive bearing forces are transferred to the glass and no tension is transmitted from the bolt to the glass. This in turn implies that a non-linear finite element analysis is required.
- Tetrahedral elements should not be used.
- The mesh density should endeavour to match the expected stress concentrations i. e. the density should be large around the hole (minimum of 32 elements) and gradually reduce away from the hole.
- Conical bore holes have to be modelled with solid 3-dimensional elements.
- Cylindrical bore holes may be modelled either with solid 3-dimensional elements or 2-dimensional shell elements.
- Any metal plate and the intermediate materials must be modelled with solid elements.

- A convergence analysis of the finite element model should be carried out to ensure that the results are accurate and that they are not overly sensitive to user defined parameters such as mesh density, non-linear convergence criteria etc.

Once the stress distribution is obtained it is then possible to adopt the complex but flexible design method described in Chapter 6 or the more conservative, but simpler allowable stress based methods described in Section 4.2. In the latter case, it is common practise to adopt the tempering stress around the bolt hole as the glass strength and to ignore the inherent glass strength. This is a conservative, yet simplified approach as it eliminates complex considerations of surface condition, stress history and area effects.

Material selection and detailing are other key considerations in the design of bolted connections in glass. As far as the glass is concerned it recommended to use either HSG or FTG, as the strength of ANG in the bore hole area is very poor. Furthermore, when laminated heat-treated glass is used there is often a misalignment in the hole region, consequently resin or injection mortar is preferred to hard bushings as the latter are unable to create an homogeneous load distribution in all glass layers.

Through bolt connection

The earliest and generally strongest type of bolted connection is the through bolt connection where the connection is subjected to in-plane tension or compression which is translated as shear in the bolts. This type of connection is derived directly from steel and timber construction and is particularly useful in glass as it can provide structural continuity between separate glass elements that are limited in size due to the manufacturing process. This type of connection may therefore be used in splices (e. g. spliced beams, spliced fins) to construct large structural assemblies.

The through bolt connection in *Figure 7.6* is constructed by drilling a hole in the members to be connected and inserting a bolt through the hole so that it transfers the forces across the joint. The bolt is subjected to shear forces, whereas the connected members are locally subjected to high bearing stresses which are evident as compressive stresses at the location of contact with the bolt and tension stresses on the sides of the bolt hole with a peak tensile stress normal to the point of contact. The latter tension stresses are due to the elongation of the hole in the direction of the applied force.

Point supports

Point supports are essentially bolted connections that are used in glass-to-glass connections or to connect glass to a subframe without creating a lap joint as discussed above. The removal of the lap joint has the benefit of reducing the visual impact of the connection and most point supports have evolved further in this regard by having a countersunk bolt that eliminates all protrusions beyond one surface of the glass.

Typical point supports are not suitable for the transfer of high in-plane loads into the glass pane except the dead load of the glass itself in case of vertical glazing. For the transfer of high in-plane loads, it is preferable to use through bolt fixings (see above).

Point fixing must often cater to some degree of lateral loading on the glass panels. The original point fixings provided a rotationally stiff connection to the subframe, which often resulted in higher stresses in the bolt hole area. Later versions provided a stiff connection that endeavoured to match the stiffness of the glass thus reducing the stress concentrations from the rotational restraint. The more recent point supports include a ball and socket joint, known as articulated bolts, which allow for free rotation of the panel.

To allow easy assembly and to avoid unfavourable in-plane constraints (e. g. due to temperature), the point-support pins should be tightened carefully (i. e. torque screw moment < 30 Nm) and fixed into slotted/oversize holes in the sub-structure with suitable low-friction interlayers (e. g. teflon) as shown in *Figure 7.8*. Point supported glass should have a minimum thickness of 8 mm and the distance of holes to the glass edges must not be less than 2.5 times the glass thickness [202].

In addition to the parameters listed in Section 7.2.3, the stress distribution around the holes of a point supported connection is also influenced by:

- The position of the point fixing in the glass panel.
- The type of point support i. e. rotationally stiff, flexible or fully articulated.
- The geometry of the bore hole e. g. cylindrical bore holes for points fixings with steel disks on each glass surface or conical holes for point supports with countersunk bolts.
- The applied torque on the fixing bolt.

Recommendations for scheme design

Well made through bolt connections in good quality FTG should be able to resist a bearing load of 0.7 kN per mm of glass thickness [277]. However, it is advisable to carry out more detailed analysis as discussed in Section 7.2.3.

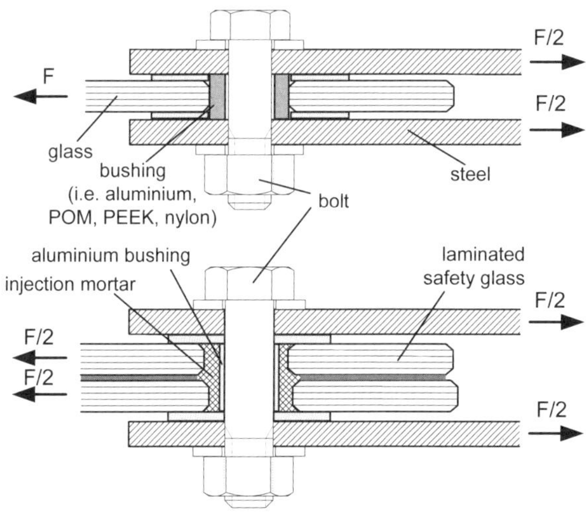

Figure 7.6:
Example of a through bolt connection; top: monolithic glass, bottom: laminated safety glass.

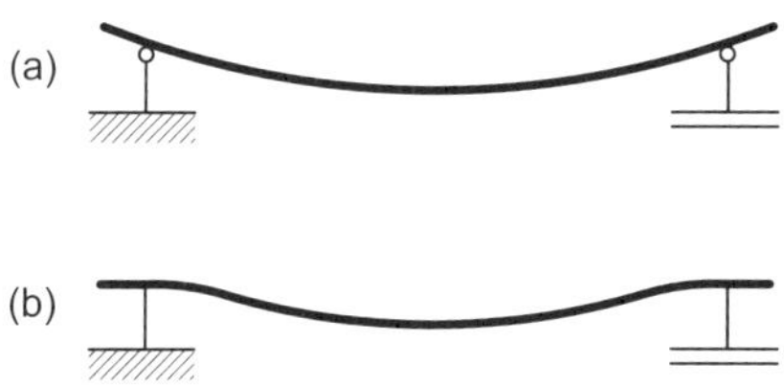

Figure 7.7:
Point support type: (a) with a free rotation of the panel, (b) with a stiff connection between point support and glass panel.

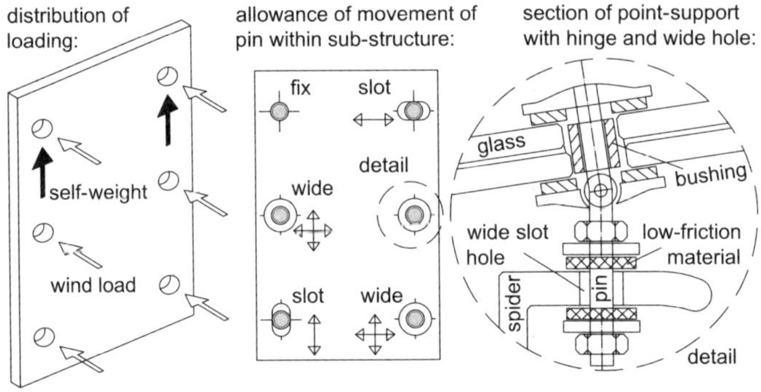

Figure 7.8: Example of a point-supported glazing panel and its support conditions (sub-structure).

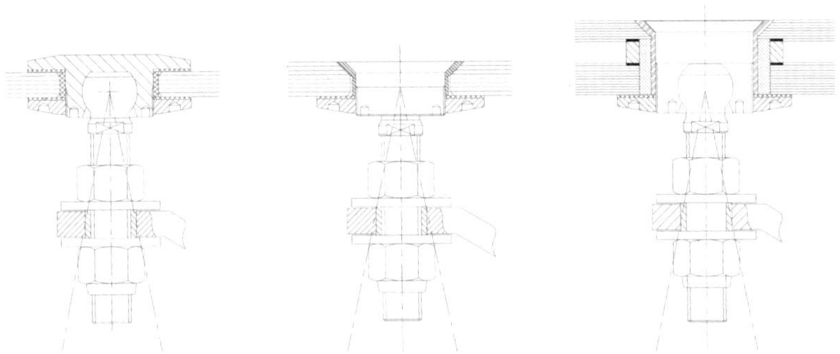

Figure 7.9: Example of point supports with cylindrical hole (left), conical hole monolithic glass (center) and conical hole with insulated glass unit (right).

The design charts shown in *Figures 7.10, 7.11* and *7.12* are reproduced from Overend and Maniatis [267] and provide useful preliminary sizing for horizontal and vertical point supported glass panels fitted with articulated bolts.

7.3 Glued connections

This text has been compiled in collaboration with the following experts:
Dr. Lucio BLANDINI, Christoph HAAS, Prof. Dr.-Ing. Werner SOBEK, Dr. Frank WELLERSHOFF

7.3.1 General

Most engineers are relatively unfamiliar with adhesive technology and terminology. The aim of this section is therefore to briefly introduce the general principles of adhesive bonding. Further detailed information on structural adhesives is provided in specialized publications on the subject, such as [5, 220].

Glued connections provide the opportunity to distribute the loads arising from the connections in a more uniform manner when compared to bolted connections. This is clearly an advantage in glass connections, which because of the brittle nature of the material are sensitive to stress concentrations. Another advantage of adhesive connections

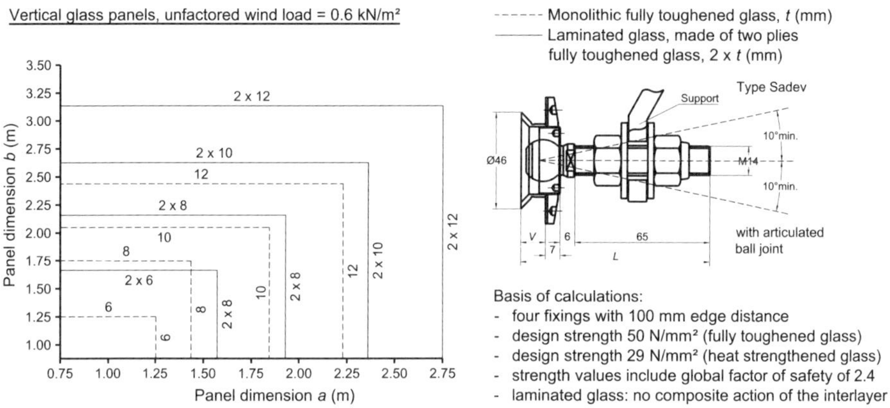

Figure 7.10: Design chart for vertical point fixed glass panels with $W_k = 0.6\,\text{MPa}$.

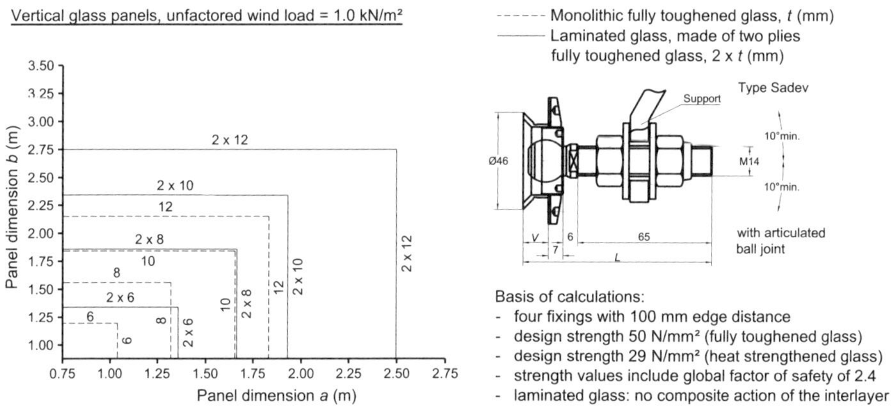

Figure 7.11: Design chart for vertical point fixed glass panels with $W_k = 1.0\,\text{MPa}$.

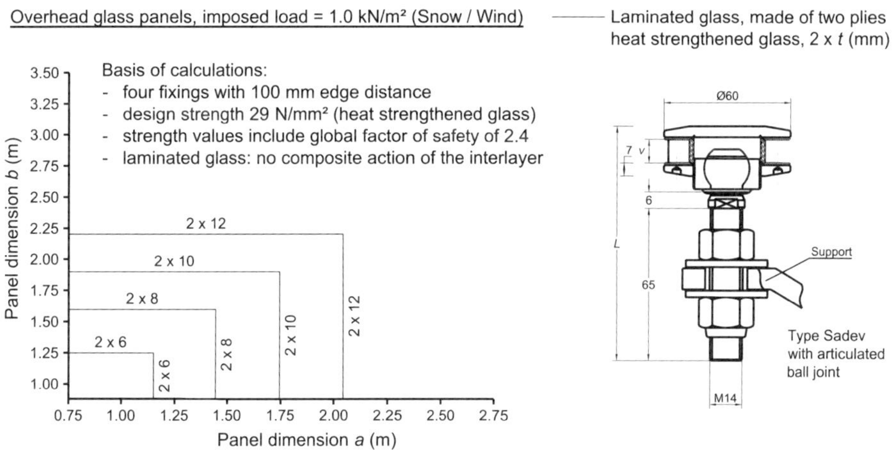

Figure 7.12: Design chart for overhead point fixed glass panels.

in glass is that it provides a surface that is flat and easy to clean, thereby making glued connections relatively easy to construct and eliminating in many circumstances the need for pretreatment of the glass surface. Generally, two types of glued connections are used for glass applications:

- soft elastic adhesive connections (i. e. structural-silicone-sealant connections),
- rigid adhesive connection (i. e. acrylic adhesives, epoxy adhesives and polyester resin).

Adhesives are polymer materials that consist of simple monomer units recurrently chained to macromolecules. The atoms in each macromolecule are chemically bonded and the macromolecules are physically or chemically bonded to each other and intertwining is inevitable (*Figure 7.13*).

Polymers can be classified according to their thermomechanical properties that are controlled by the molecular structure. Silicone adhesives and one-component polyurethanes are examples of typical elastomers. Two-component polyurethanes are more likely thermosets with a high cross-link density.

Thermoplastics Relatively weak intermolecular forces hold molecules together in a thermoplastic, so that the material softens when exposed to heat, but returns to its original condition when cooled. Thermoplastic polymers can be repeatedly softened by heating and then solidified by cooling - a process similar to the repeated melting and cooling of metals. Most linear and slightly branched polymers are thermoplastics. All the major thermoplastics are produced by chain polymerization.

Elastomers Elastomers are rubbery polymers that can be stretched easily to several times their unstretched length and which rapidly return to their original dimensions when the applied stress is removed. Elastomers are cross-linked, but have a low cross-link density. The polymer chains still have some freedom to move, but are prevented from permanently moving relative to each other by the cross-links.

Figure 7.13:
Molecular structure of polymers.

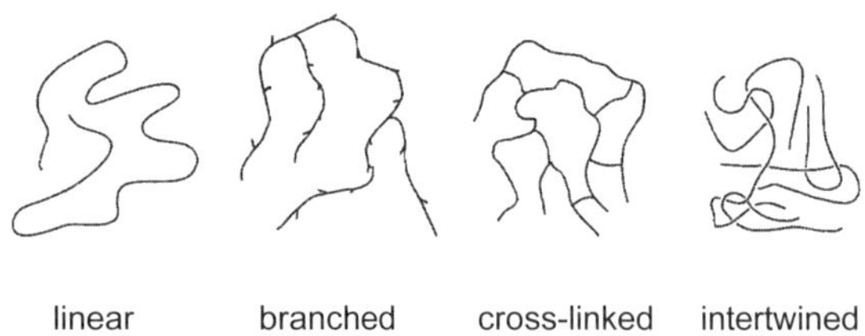

Figure 7.14:
Classification of polymer adhesives [188].

Polymer adhesives

Thermoplastics	Elastomers	Thermosets
Linear or branched chains	Long cross-linked polymer chains	High cross-linked polymer chains
PVB	Silicone (inorganic) Polyurethane (organic)	Acrylic adhesive Polyester resin, Epoxy

Thermosets A thermosetting plastic solidifies or 'sets' irreversibly when heated and further heating cannot reshape this material. Thermosets consist of 3-dimensional networked polymers with a high degree of cross-linking between polymer chains. The cross-linking restricts the motion of the chains and leads to a rigid material.

Under external forces three different deformation types, which have to be superimposed, could be identified:

A. Spontaneous elastic deformation (spontaneous reversible) due to changed valence bond angles of atoms in chemical bonding.
B. Time dependent viscoelastic deformation (time dependent reversible) due to stretched molecular chains.
C. Time dependent viscoplastic deformation (time depending irreversible) due to movement of molecular chains.

The ratio of the deformation type A, B or C depends on the molecular structure of the adhesive. For low cross-linked polymers the deformation types B and C are more important than deformation type A.

Strong atomic bridges between the adhesive and the glass surface result in a strong adhesive joint. The glass surface consists of silicon atoms saturated with OH-groups and some metal ionic (i. e. Na) as shown in *Figure 7.15* and it is therefore desirable to establish strong Si-O-Si bonds between the glass surface and the adhesive. This may be achieved by applying a silanized primer to the surface and an adhesive with a compatible molecular structure. The primer has the dual function of providing a reactive group for the glass surface in addition to a reactive group for the adhesive [285].

The atomic bonding happens in three steps:

1. Hydrolysis of the silane in the primer to a silanol is enabled by the humidity on the glass surface.
2. Hydrogen bonds arise between the OH-molecules of the silanol and the glass surface.
3. By splitting of water, some hydrogen bonds change into chemical Si-O-Si bonds.

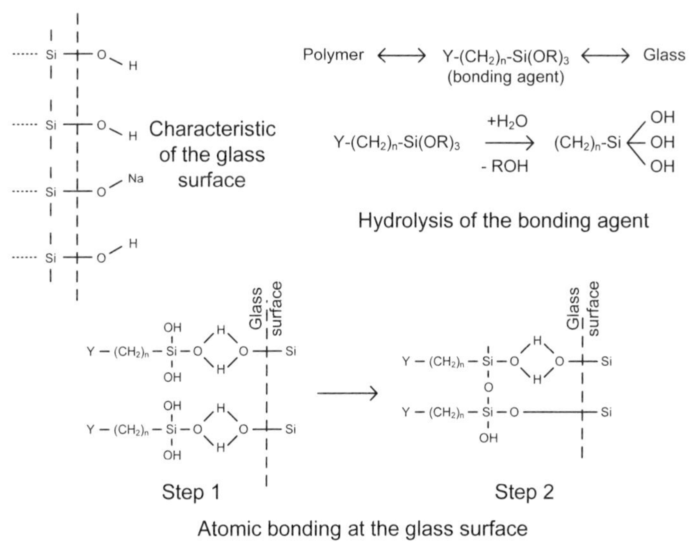

Figure 7.15: Hydrolysis of the bonding agent and atomic bonding at the glass surface.

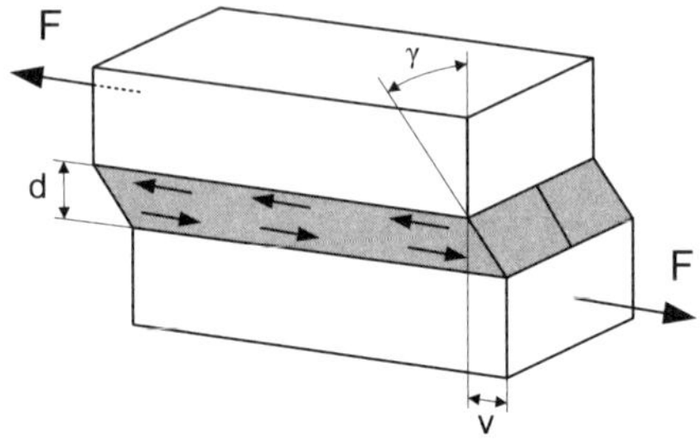

Figure 7.16:
Adhesive deformation under shear force.

Mechanical behaviour of adhesives

Behaviour under short-term loads and small strain The deformation of an adhesive layer between two bonded elements is shown in *Figure 7.16*. The relation between the shear modulus G, the shear deformation $\tan\gamma$ and the shear stress τ are defined as

$$\tan\gamma = \frac{v}{d} \tag{7.1}$$

$$G = \frac{\tau}{\tan\gamma} \tag{7.2}$$

where v is the shear displacement and d is the adhesive thickness.

Behaviour under long term loads and small strain Long-term loads are those due to self weights. A common approximation for the time-dependent elasticity I is

$$I = \frac{\tan\gamma}{\tau} = B \cdot t^{\alpha} \tag{7.3}$$

where B and α are material parameters. The change in shear deformation over time is best expressed by a double logarithmic plot(*Figure 7.17*). Three different regions may be defined, each representing varying degrees of creep:

- I Primary region: The creeping is provoked by the stretching of the molecular chains.
- II Secondary region: The creeping is provoked by sliding of the molecular chains. The lost physical bondings between the molecular chains and the gained physical bondings are in balance.
- III Tertiary creeping: The lost physical bonding becomes prevalent and the connection breaks.

The parameters γ_{I_0}, γ_{II_0} and γ_{III_0} are the shear strains at the beginning of the deformation in regions I, II and III. The time intervals in the respective regions are Δt_{I}, Δt_{II} and Δt_{III}. The ultimate limit values are γ_{B} and t_{B}. For design purposes, γ_{III_0} should not be reached because the failure of the connection is initialized at this level.

7.3.2 Structural silicone sealant connections

Structural silicone sealants were originally applied to bond glass panes to aluminum subframes in curtain wall facades of high rise buildings (structural sealant glazing systems, SSGS). However, structural silicones are increasingly being used to achieve soft structural connections between glass and aluminum or stainless steel or between glass and glass.

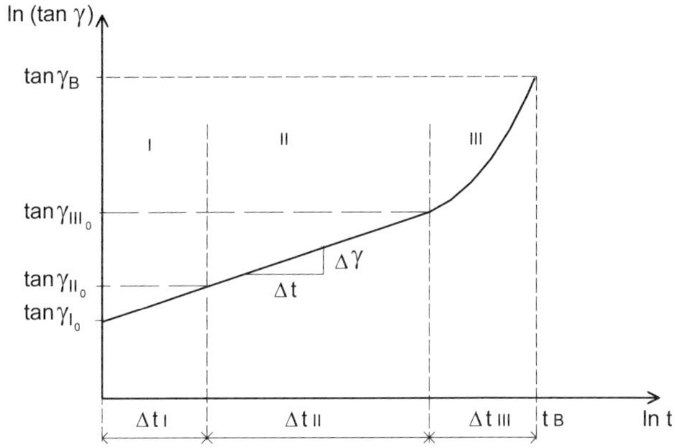

Figure 7.17:
Time-dependent load carrying behaviour under long term loads.

These connections are employed to create 'transparent' glass structures where mechanical fixings are replaced by structural silicone joints (e. g. glass corners, glass fins bonded to fully glazed vertical glass). Two different types of structural silicone sealants are available:

- *One-component silicones* start curing as soon as they come into contact with moisture in the air. Optimum conditions for application are 24 °C at minimum 50% relative humidity. The diffusion-controlled curing process imposes practical limits on the geometry of the seal: recommended thickness > 6 mm, maximum width < 20 mm. The ratio of joint thickness to joint width must be at least 1 : 1 but no more than 1 : 3. A ratio of 1 : 2 is ideal. Depending on the thickness, curing times up to 3 weeks have to be considered. If the seal is too thick, the interior parts may never cure completely.
- *Two-component silicones* are cured by the polymerization reaction that is triggered by the mixing of the two components that consist of a base compound (about 90% by volume) and a catalyst (about 10% by volume). The curing does not require outside chemical components. Diffusion lengths among the two components are very small and curing will progress relatively quickly (curing time less than 3 days), homogeneous and independent of the joint size. The recommended minimum thickness is 6 mm, the maximum width is 50 mm. Depending on manufacturer recommendations or design codes a maximum ratio of joint thickness to joint width of 1 : 4 is allowed. Proper mixing is very important and must be checked frequently during application. Therefore application of two-component silicones on the building site is generally problematic and should be avoided.

Material properties may differ from one manufacturer to another. Common values are given in *Table 7.1*. In small-scale short-term laboratory testing structural silicone sealants typically achieve tensile strengths of around 0.8 to 1.8 MPa for dynamic loading, depending on the temperature. However, allowable stresses for wind loads are usually much lower. Creep is initiated under long-term loading stresses equivalent to roughly 10% of the short-term strength. Long-term strains in excess of creep levels will lead to relaxation. This will reduce the stresses but no failure will occur, as long as permissible strains are not exceeded. Structural silicone joints are normally designed in terms of allowable stresses (*Table 7.1*), which are in turn based on the ultimate strength and a safety factor of 6 for short-term loads and that of 60 for long-term loads (e. g. European Technical Approval Guideline (ETAG) No. 002 [163]). The allowable strain range under shear deformation due to differential thermal dilatation is ±12.5% of the ultimate strains. At this relatively low level of strains, it is sensible to assume elastic behaviour.

The very low modulus of elasticity constitutes both an advantage and a disadvantage. On one hand it reduces stress concentrations, but on the other hand structural silicone sealants are not suitable to transfer high shear forces required for built-up sections of glass (e. g. T-section, H-section).

When used in combination with laminated safety glass, structural silicone sealants show good behaviour in case of protective glazing (i. e. where facades are subjected to impact or blast loads). This is due to the soft material behaviour that has a capacity to absorb high amounts of energy.

The design of a structural silicone joint must allow for sufficient load-carrying strength in order to transfer the applied loads. At the same time the allowable strain of the silicone sealant must not be exceeded. The maximum strain is particularly critical if two materials with different coefficients of thermal expansion are bonded together. For this reason, joint geometries with adhesion to three surfaces such as L-shaped joints have small displacement capacities (*Figure 7.18*) and should be avoided, for example, when the glass panel is glued with all edges onto an aluminium frame in a curtain wall facade element.

The influence of SSG joints on the load carrying behaviour of glass panels is studied in [332]. In case of a high glass edge rotation due to glass deformations in combination with large structural sealant joints the resulting additional tensile stress in the joint has to be taken into account. Due to the Poisson's ratio μ of nearly 0.5 the stiffness of a structural sealant joint depends strongly on the geometry of the joint–this explains the recommended width to thickness ratios given by manufacturers. For structural applications, i. e. the fixing of glass fins or any in-plane load introduction, different joint configurations are shown in *Figure 7.19*. The load carrying behaviour depends on the ratio between face side bond length and lateral bond length. The load is mainly transferred over the face side

Table 7.1: Typical material properties of structural silicone sealants (manufacturers data).

Property	Symbol	Unit	Value
Allowable tensile stress, short-term loads	$\sigma_{\text{all,short}}$	MPa	0.14
Allowable tensile stress, long term loads	$\sigma_{\text{all,long}}$	MPa	0.014
Allowable shear stress, short-term loads	$\tau_{\text{all,short}}$	MPa	0.070–0.128
Allowable shear stress, long term loads	$\tau_{\text{all,long}}$	MPa	0.007–0.011
Young's modulus of elasticity, short-term loads	E_{short}	MPa	1.0–2.5
Maximum allowable strain [215]	ε_{all}	–	±12.5%
Poisson's ratio	ν	–	0.49

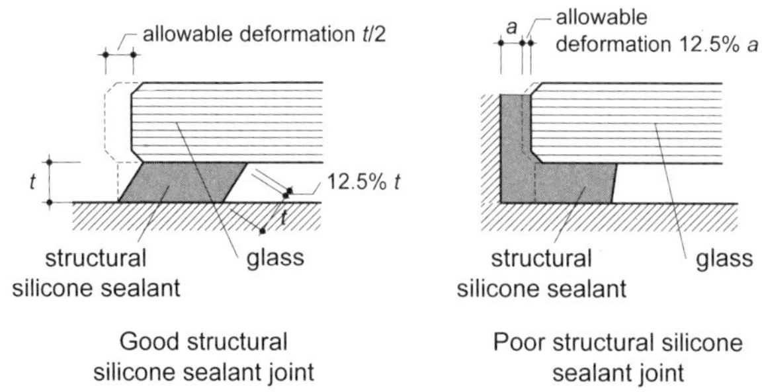

Figure 7.18: Example of a good (two sided) and a poor (L-shaped) structural silicone sealant joint.

joint which is subjected to tension. Due to the lateral elongation restraint in the U-channel, the lateral bond length has only a small influence on the joint stiffness. Approved design methods for these types of connections are not yet available and research is still ongoing [55, 188].

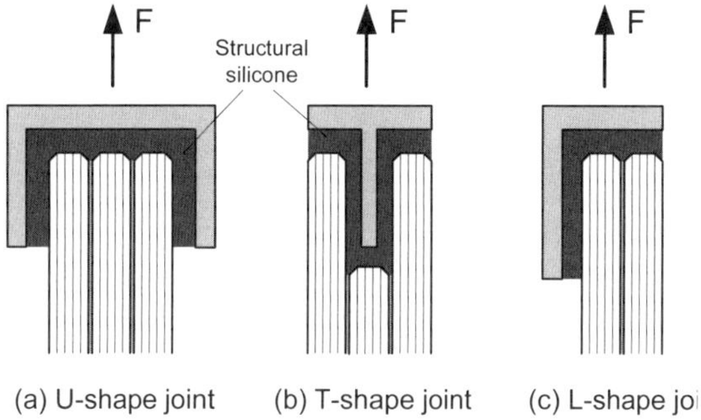

Figure 7.19:
Different structural silicone joints.

In some countries such as Germany or France structural silicone sealant cannot be used to carry permanent loads (i. e. dead load) and SSGS facades require additional mechanical fixings to prevent the glass from falling once the silicone fails.

Furthermore, the chemical compatibility of all materials in contact with the silicone must be ensured in order to ensure long-term performance and prevent damage. The compatibility of any material that the structural silicone adhesive comes in contact with (e. g. gaskets, spacers, backer materials, setting blocks) has to be approved by the manufacturer or should be tested in the laboratory. EPDM, neoprene, bitumen, asphalt and other organic-based membranes, coatings and gaskets often cause discolouration of light-coloured silicone sealants. These materials are often approved for incidental contact with the structural silicone but are not approved for full contact as a structural spacer material [7].

The design values provided by manufacturers are based on the assumption that the sealant fails due to loss of cohesion (i. e. failure within the silicone) rather than adhesion (i. e. failure at the silicone-adherend interface). Adhesion quality is mainly affected by the surface quality of the connected materials (adherends). Flat surfaces such as glass provide the best conditions while surfaces with pores are unfavourable as they only allow adhesion to the local peaks in the material. Aluminum, anodized aluminum, stainless steel (not brushed or with satin finish) and some powder coatings offer good conditions for adhesion. Some glass coatings (e. g. most self-cleaning coatings) are not suitable with structural silicone. All surfaces have to be primed prior to the silicone application [90].

In Europe, the application of SSGS are regulated, details of which can be found in [53, 75, 163]. A European Technical Approval (ETA) for any silicone used in structural applications is needed. EOTA 1998 [163] defines permissible stresses for loading in dynamic (short-term) tension, in dynamic shear and in permanent shear (but not for permanent tension). For dynamic tension it requires that the 5%-fractile value of the strengths measured on small scale tests must exceed the permissible stress by a factor of 6. For permanent shear a minimum creep factor of 10 is defined. Therefore permissible stresses for permanent shear loads are usually 10 to 15 times lower than that for dynamic shear loads. The design method given in EOTA 1998 [163] is limited to four-side supported glass (SSGS glued or mechanical fixed) panels with a linear SSGS joint over the entire glass edge. This quite rough design approach does not take into account the stiffness of the supporting frame or the non-linear stress distribution along the glass edge in the SSGS joint due the deformation of the glass panel. An application of EOTA guidelines [163] for any structural silicone sealant connection is therefore not useful. EOTA requires all SSGS connections to be made in the factory rather than on site. This is because proper execution of an SSGS joint requires a controlled climate and clean surroundings. However, in all-glass structures some SSGS joints may have to be applied on-site. This will require special measures to ensure a proper environment and even more stringent quality assurance procedures. Even so, it should be noted that the structural quality of an SSGS joint cannot be tested non-destructively.

In the USA, SSGS applications are regulated by [1, 15]. The design principle is similar to the European approach. Dow Corning provides a detailed design guide and examples on detailing [89].

7.3.3 Rigid adhesive connections

The search for connection elements with a minimal visual impact has led to intense research in the field of rigid glued connections. Structural silicone is the only adhesive product with a proven track record in glass architecture; however this product is unsuitable for minimal discrete adhesive joints as it is neither strong nor stiff enough for this application. Epoxies and acrylics, which have been used successfully for decades in the aeronautical and automotive industries, are the most promising stiff adhesives for glass construction. However, their performance is largely untested and there are a number of challenges in transferring the technology from other industries to glass construction [321].

Parameters affecting performance

There are many aspects to be considered in the design of a rigid adhesive joint, including the selection of a suitable adhesive in terms of chemical compatibility, the geometry of the bonded area, the temperature range in which the adhesive must perform, the durability of the adhesive joint as well as issues relating to workmanship and curing.

Some of the first choices to be made in designing an adhesive joint are those relating to the *geometry of the adhesive bond*. The thickness of the adhesive layer is a primary consideration in this respect. At this stage, it is important to distinguish between *contact adhesives* that require a small adhesive thickness often below 1 mm and *gap-filling adhesives* that are able to perform at thicknesses in excess of 5 mm. Although annealed glass is a relatively flat product, heat treating the glass causes roller wave distortions, and fixing two or more pieces of glass will sometimes require further assembly tolerances. With tolerances in excess of 1 mm it is recommended to use epoxy-based adhesives which have gap filling properties. With lower tolerances it may be possible to use contact adhesives such as acrylic-based adhesives.

UV-curing acrylics have already been used successfully in the furniture industry to assemble pieces made entirely of glass and have been shown to achieve a good joint. Unfortunately, they do not cure effectively in thick layers, because the UV radiation is not able to reach all the monomers to activate the polymerization. Such adhesives are promising for their use with glass because they are transparent and once cured they are resistant to UV radiation. One of the most interesting applications in architecture to date is the 2001 renovation of the Austrian IBM head office in Vienna: here for the first time a stiff adhesive, instead of structural silicon, is used to structurally bond the façade to the underlying metal structure.

The perimeter shape of the adhesive joint is also an important geometrical consideration. One of the disadvantages in using stiff adhesives is their limited capacity to redistribute stress concentrations and to absorb deformation. It is therefore necessary to avoid geometrical singularities and sharp edges of the adherents: these lead to stress concentrations [3] and in some cases can be avoided by rounding the edges. If a finite element calculation of the joint is pursued, it is important to include such rounding in the geometrical model, since FE-models could overestimate the stress concentration around points of singularity.

Another aspect to be considered is the *temperature range* that the connection has to withstand during its service life. If the temperature is above the glass transition temperature of the adhesive T_g, the chain segments of a macromolecule can move, thus leading to a reduction in stiffness and strength. Below T_g such movements are frozen. The glass transition temperature depends on the chemical composition and on the cross-linking rate of the adhesive. Epoxies often have a T_g higher than acrylics and are therefore generally more suitable for application at higher temperatures. Both the short-term and the long term joint behaviours are influenced by the T_g: the value of the force the joint carries is generally lower when approaching T_g and the viscous component of the deformations increases as the temperature gets closer to T_g [292, 295]. When T_g is reached, the change in the fundamental quantities is not abrupt, but gradual [4]. Moreover, the adhesion forces are lower at high temperatures, so the same product may exhibit a cohesion failure at low temperatures and an adhesion failure at high temperatures [43].

The need for a surface treatment to improve adhesion on glass depends on the kind of adhesive. The glass surface to be bonded can be treated by means of mechanical or chemical processes or with the use of primers, which may provide a more receptive layer to the adhesive. Initial information on certain products as well as on certain treatments is available, but this field needs further investigation [241].

The last important aspect to be taken into account is *durability* [228]. There is a lack of research on the long term performance of glass adhesives [241], so the following information draws from the experience gained in the field of metal adherents. The influence of service time on adhesives depends on their chemical composition and on their cross-linking rate [187]. It has been found that the mechanical properties of an adhesive joint, which depends on the adhesive layer itself as well as on the interface adhesive adherents, may deteriorate upon exposure to its service environment: water, in liquid or vapour form, is the most hostile environment for structural adhesive joints that is commonly encountered. The influence of water on the adhesive is generally reversible, so that any deterioration is recovered upon drying. All polymers absorb greater quantities of water when their temperature is above T_g , so that rubbery materials tend to show greater water absorption than rigid adhesives [251]. Other parameters affecting environmental durability are temperature, stress rate and distribution in the adhesive layer as well as surface characteristics and pretreatment of the adherents. Accelerated weather tests have been set up in laboratories to accurately model the complex deterioration process within a reasonable range of time [87, 128, 144, 145, 158, 159, 160, 161]. These are obtained by increasing the presence of certain agents above their natural rates: UV radiation, moisture or water and temperature. Specimens may also be immersed in acids or salt solutions. The results obtained should preferably be compared to published information on natural weathering behaviour.

In addition to the influence of aging, the effect of each of the previously described parameters has to be investigated by means of tests that could be carried out on bulk specimens or on the whole joint. Similar tests are carried out for overlapping metal joints [19, 20, 86] so that in certain cases it is possible to adapt the existing tests for glass applications.

In general, specific research on the use of glass and stiff adhesives has focused on assembling all-glass systems or on developing mixed structures. The former includes proposals for glass adhesive T-beams composed of two glass panes [282], for a glass cruciform column composed of three pieces of glass [264] and for a glass shell (*Figure 7.20*) assembled by means of adhesive butt joints [43]. The latter includes research on composite beams made of a wood frame glued onto glass which has a stiffening function [191, 233, 260] and glass-fibre-reinforced plastic profiles glued on glass plates [229]. This innovative research provides a glimpse of the opportunities offered by stiff adhesives, but it is important to note that there is still a lack of understanding of the basics in glass adhesion. therefore a substantial amount of further research is required in this field for glass adhesion to become an accepted and mainstream form of construction.

Figure 7.20:
Example of a glass shell with butt adhesive joints. (designed and built by Sobek and Blandini, University of Stuttgart, Germany)

Limit state design

According to the limit state analysis, the safety factors used in the design of adhesive joints must take into account the uncertainties associated with the fabrication and analysis of the joint (effects of workmanship, uncertainties concerning the assumed stress distribution in the joint) and the changes in material with time. Suggestions for the values of the safety factors are given in the EUROCOMP Design Code and Handbook [68] (*Table 7.2*). In this proposal it is assumed that the adhesion to glass is adequate, so that only the behaviour of the adhesive itself is considered. With this premise, the material safety factor is calculated as

$$\gamma_{\mathrm{m}} = \prod \gamma_{\mathrm{mi}} \geq \begin{cases} 2 & \text{for connections designed by testing,} \\ 4 & \text{for connections subjected to long-term loading.} \end{cases} \tag{7.4}$$

Table 7.2: Recommended values for partial safety factors to be applied to adhesive properties [68].

Source of adhesive properties	γ_{m1}
Typical or textbook values	1.5
Values obtained by testing	1.25
Source of adhesive properties	γ_{m2}
Manual application, no adhesive thickness control	1.5
Manual application, adhesive thickness control	1.25
Established application procedure with repeatable and controlled process parameters	1.0
Type of loading	γ_{m3}
Long-term loading	1.5
Short-term loading	1.0
Environmental conditions	γ_{m4}
Service conditions outside test conditions	2.0
Adhesive properties determined for the service conditions	1.0
Fatigue loading	γ_{m5}
Adhesive subjcected to significant fatigue loading	1.5–3.0
Loading basically static	1.0

For the design of adhesive joints in case of fire it will be necessary to carry out a heat flow analysis and determine the capacity of the joint under the design temperature [220]. However, where fire is a major design consideration, a pure adhesive joint will not be appropriate, unless the adhesive can be effectively insulated.
Wellershoff [338] suggests the following approach for the design of glued connections

$$\frac{\tau_{\mathrm{Ed}}}{\tau_{\mathrm{Rd}}} = \frac{\tau_{\mathrm{Ek}} \cdot \gamma_{\mathrm{F}}}{\dfrac{\tau_{\mathrm{Rk}} \cdot f_{\mathrm{T,t}}}{\gamma_{\mathrm{M}}}} \leq 1 \tag{7.5}$$

where τ_{Ek} represents the characteristic value of the shear stress, τ_{Rk} the characteristic value of short-term shear strength, $f_{\mathrm{T,t}}$ a reduction factor which is a function of the material temperature T and the load duration t (represented schematically in *Figure 7.21*), γ_{F} is load safety factor (according to national code) and γ_{M} is the material safety factor (according to national code).

Experimental studies have been carried out on the time- and temperature- dependent behaviour of adhesives that are used in glass construction, and diagrams have been developed for the reduction factor $f_{\mathrm{T,t}}$ [338]. Additionally creep effects under a constant material temperature T may be quantified using Equation (7.3) and Equation (7.5)

$$f_{\mathrm{T,III0}} = f_{\mathrm{T,I0}} \cdot \frac{I_{\mathrm{T,I0}}}{I_{\mathrm{T,III0}}} = f_{\mathrm{T,I0}} \cdot \frac{B_{\mathrm{T}} \cdot t_{\mathrm{T,I0}}{}^{\alpha}}{B_{T} \cdot t_{\mathrm{T,III0}}{}^{\alpha}} = f_{\mathrm{T,I0}} \cdot \left(\frac{0,1}{t_{\mathrm{T,III0}}}\right)^{\alpha} \tag{7.6}$$

where $I_{\mathrm{T,I0}}$ is the creep resistance at the beginning of deformation region I with temperature T and $I_{\mathrm{T,III0}}$ is the creep resistance at the beginning of deformation region III with temperature T. For other stresses in glued connections the method is similar.

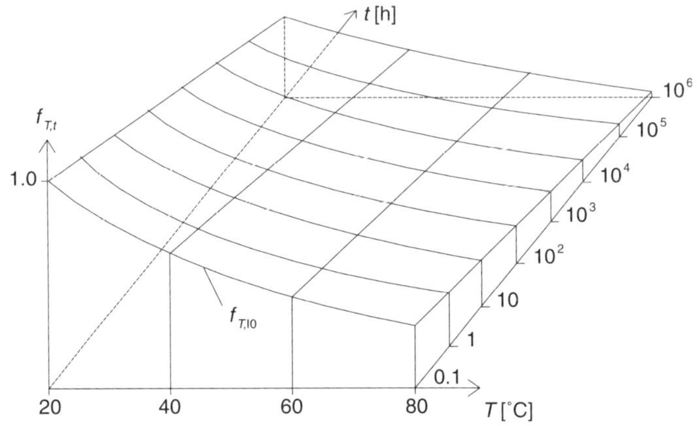

Figure 7.21: Schematic representation of the reduction factor $f_{T,t}$.

7.4 Recent developments and trends

This text has been compiled in collaboration with the following experts:
Prof. Dr. Mick EEKHOUT, Dr. Jürgen NEUGEBAUER, Prof. Dr. Geralt SIEBERT, Ronald VISSER

It is evident that there is still a lack of knowledge concerning glass connections and the current connections are largely an adaptation of pre-existing connections in steel and timber construction. As structural glass becomes even more popular it is necessary to devise new and more suitable connections and develop design methods that enable the structural engineer to perform simple yet accurate calculations. The objective of new connection types has to be focused mainly on:

- The improvement of the load carrying behaviour.
- The improvement of the load bearing capacity after partial or total failure of a glass member.
- The development of new connection systems that are more suitable for the brittle material behaviour of glass.
- The development of connections that provide some structural redundancy, thereby increasing the safety of glass structures.
- The combination of various connection types in order to compensate unfavourable properties of one connection type by favourable properties of another.

7.4.1 Increasing the post-breakage structural capacity with fabric embeds

A new concept that increases the post-breakage structural capacity of laminated safety glass in overhead glazing is to locally reinforce the PVB interlayer with fabric. Neugebauer [258] provides some ideas on how this fabric might be employed for different support conditions. In simply supported overhead glazing with glazing beads the fabric is embedded into the PVB-interlayer between the glass panes near the edges, and the fabric protrudes from the edge of the glass (*Figure 7.22a*). The overlapping fabric is cast into a plastic fastening border, or is welded to a metal profile that is screwed to the substructure. In the event that the laminated glass is broken, the fabric prevents the glass from falling from the frame.

This reinforcement concept may also be applied to reduce the risk of post-breakage tear out in point supported overhead glazing (*Figure 7.22b*). In this case, a circular piece

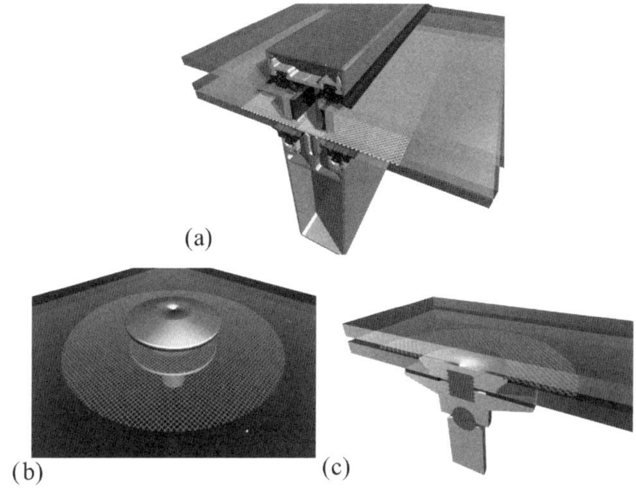

Figure 7.22:
Fabric reinforcement of laminated safety glass supports.

of mesh is embedded into the PVB-interlayer around the bolt hole in the glass. The fabric is cast into a plastic hollow shaft or welded to a metal hollow shaft which is then fixed to the point fitting. In undercut point fittings a conical hole is drilled only in one glass pane of the laminated safety glass. A circular fabric may be embedded into the interlayer between the glass panes and is then welded to the special conical glass fitting (*Figure 7.22c*). The whole system, consisting of fabric and conical undercut glass fitting, should be assembled during the lamination process.

7.4.2 Increasing the post-breakage structural capacity with new geometries

New geometries of point supports may increase the post-breakage structural capacity. In case of point supported glass elements, this capacity depends on the type of point fixture. In some countries, countersunk fixings are not allowed for overhead glazing as they show a very poor post-breakage load bearing capacity. Due to the small contact area between countersunk and glass, there is a risk of the panel tearing out of the countersunk bolt and fall down. This problem might be mitigated by using standard bolts where the raised head provides a bigger contact area. Usually, with raised head fixtures, the through hole is between 16 and 30 mm. The plate diameter is between 50 and 80 mm, thus offering an ample amount of contact area for support even after breakage. Nevertheless architectural demand for flush glass surfaces means that despite their better post-breakage performance such connections are not very popular.

A solution might be a safe countersunk (SCS) fixing [317] which – at first glance – looks like any ordinary countersunk point fixture (*Figure 7.23*). It has an aluminium countersunk sleeve, a stainless steel fastening head, a plastic washer, a stainless steel body plate, a threaded bolt, nuts and washers. The main difference is a special bore hole geometry that allows a countersunk for a flush design and a through hole for residual load-bearing capability. The important factor here is to have the interlayer (PVB) run all the way to the 22 mm through hole to allow a clamping of the PVB and the upper glass ply. This creates the necessary support to achieve an adequate post-breakage structural capacity.

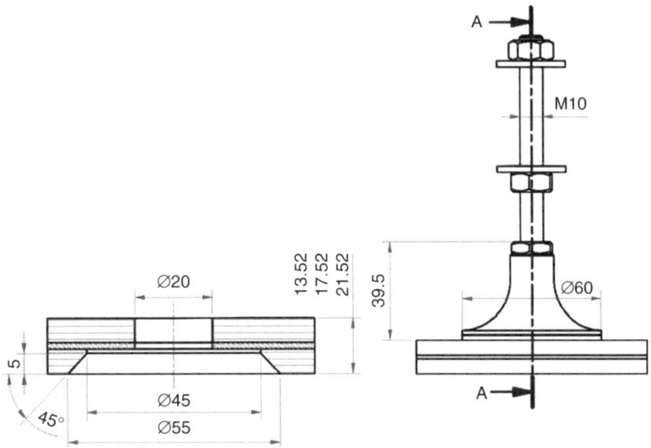

Figure 7.23:
Safe countersunk fixing with increased post-breakage structural capacity.

7.4.3 High capacity adhesive connections

Despite their popularity, bolted connections are not the ideal type of connection for glass. Firstly, glass is a brittle material, which is why local stress concentrations around the bolt cannot be reduced through stress redistribution. This makes bolted connections relatively inefficient from a structural point of view. Secondly, the surface flaws in glass caused by the drilling of holes, and the distortions of the tempering stresses around holes (cf. Section 3.6) mean that bolted connections are inducing the highest stress concentrations in the weakest possible area of the glass panel.

Research carried out by Overend [263] investigates the strength of steel-to-glass adhesive joints that eliminate the need to drill through the glass. A circular 60 mm diameter adhesive area was used and three different adhesives were tested. The best performing adhesive was an acrylic-based adhesive and achieved an average load bearing capacity of 85 kN. A series of equivalent 60 mm diameter through bolt connections was also tested for comparison, and these archived an average capacity of 29 kN. These tests showed that with the correct surface preparation and adhesive selection it is possible to provide an adhesive connection that would improve the short-term strength of bolted connections by close to 300%.

A further development in this area is the combination of a glued connection and a pretensioned bolted connection (*Figure 7.24*). This connection consists of two stainless steel adhesive discs, which are glued exactly opposite of each other on either side of a fully tempered glass plate. The discs are connected with each other by means of a stud bolt that fits tightly in the discs but goes through a clearance hole in the glass thereby bearing onto the steel discs but not the glass. The discs have an annular area for the adhesive that is clear of the relatively weak hole edges in the glass. The stud bolt is pretensioned after the adhesive has completely set. This reduces the deleterious effects caused by the peel stresses. The hole in the glass is large enough to cope with the tolerances of the glass manufacturer and to allow the bolt to bend. The adhesive has the function of an intermediate material and is being used for the transfer of the shear force between the steel discs and the glass surface and consists of a thin layer (0.1 mm) that guarantees a stiff and creep resistant joint. The pre-stressing of this connection is only possible if the discs are glued to monolithic glass. When laminated glass is required, the discs should be connected to only one of the glass layers, normally the one which is protected from weathering and vandalism.

This connection has been tested at the Faculty of Civil Engineering, Delft University of Technology. For the tests, a 19 mm fully tempered middle layer glass and stainless steel discs with a diameter of 120 mm, a thickness of 15 mm and a bolt diameter M24 have been used. The specimen failed by local overload of the glass cross-section just below the stainless steel discs. The adhesive joint survived in all the tests. The average strength of the connections was 230 kN. The stud bolt through the connection showed a visible plastic deformation before the glass failed. Such a behaviour might be used in practice as an early warning mechanism in case of overloading.

It is evident that further research is required in the promising field of high strength adhesives and glass. It must however be realized that although adhesives could provide connections that reduce stress concentrations in glass, they do not necessarily eliminate the possibility of stress concentrations. The flatness and alignment of the metal plates with respect to the glass may lead to eccentricities and stress concentrations. In some cases it may be possible to reduce stress concentrations in the glass by using metal plates that have a stiffness that is equivalent to that of the glass they are connected to.

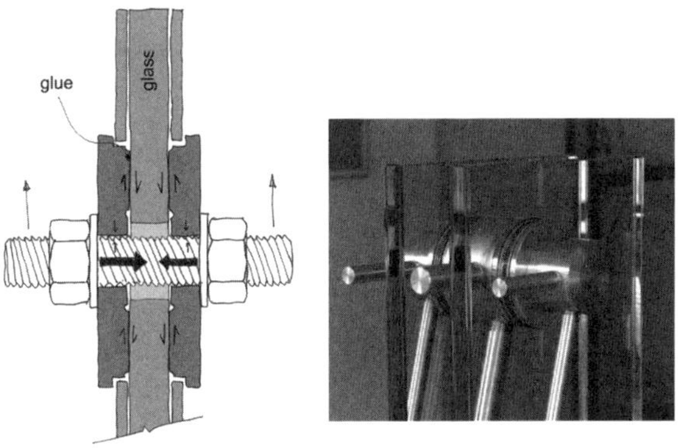

Figure 7.24:
Pretensioned adhesive connection.

Chapter

8 Special Topics

8.1 Design assisted by testing

This text has been compiled in collaboration with the following experts:
Benjamin BEER, Dr. Iris MANIATIS, Prof. Dr. Geralt SIEBERT

8.1.1 Introduction

Despite advances in the field of computational analysis, the design of complex glass structures cannot be based solely on numerical simulation. The reasons why full scale prototype testing remains an integral part of the design process of innovative glass structures, as well as the main issues that should be considered when testing glass elements, were discussed in Section 6.4.1.

Computational modelling, typically finite element models verified by rules of thumb, are required to predict the structural behaviour with an acceptable level of accuracy. The results from these calculations are often the basis for the first test prototype or specimen. Geometrical imperfections as well as tolerances should be taken into account to achieve a realistic test setup. A comparison between test results and the corresponding predicted values given by the model should be carried out. If major discrepancies are found, both the test setup and the model should be checked.

The fracture strength of heat-treated glass is the sum of the absolute value of the residual (compressive) surface stress and the inherent glass strength (see Section 3.3.2). Only the latter is influenced by subcritical crack growth and depends, therefore, on time and environmental conditions. The residual stress is constant. Consequently, results from experiments with heat-treated glass (heat-strengthened glass or fully tempered glass) in ambient conditions depend significantly less on time and environmental conditions than the results from tests on annealed glass.

General guidelines for design assisted by testing are given in the annexe of EN 1990:2002 [135]. The engineer must, however, bear in mind that this standard has not been specifically written for glass structures. Detailed reviews of the countless national standards, regional standards, building regulations and recommendations for project specific glass testing is beyond the scope of this document. *For any project, the testing procedure has to be chosen to suit the project-specific needs as well as the requirements defined by building owners, insurers and authorities.* Nevertheless, in order to provide the reader with a general idea, a few examples are discussed in the following text.

8.1.2 Post-breakage structural capacity

Project-specific testing is the common approach to ensure that a given glass assembly provides sufficient structural capacity after failure for a certain period of time. Such tests require fewer test specimens than strength tests.

The following testing procedure, which is based on a German recommendation, is often used in Western Europe for determining the post-breakage performance of a laminated glass plate that is placed horizontally and loaded laterally. It is reproduced here as one example from the many testing procedures that are currently used.

Before testing, the specimen is loaded to the greater of half the service load or $0.5\,\mathrm{kN/m^2}$. For accessible glazing, the load should be applied using 1 kN weights applied to areas of 200 mm × 200 mm. With the load still applied, all the sheets of the laminated glass panel are fractured at several locations by means of a hole punch and a hammer. The laminated glass panel is deemed to pass the test if the specimen does detach itself from the supports and no dangerous glass fragments fall down within 24 hours from testing.

8.1.3 Impact testing

Examples for railings and balustrades

- **Europe — EN 12600 [105].** The test simulates human body impact using a 50 kg mass wrapped by two rubber tires (soft pendulum test). The test is intended to classify flat glass products according to their impact resistance performance and mode of breakage.

  The test setup, including the geometry of the test frame and the impact body, is specified in detail. The impact body must have a weight of 50 ± 0.1 kg, the tyre pressure is 0.35 ± 0.02 MPa. The specimen's dimensions are 876 ± 2 mm × $1\,938 \pm 2$ mm (independent of the size of the actual component). The glass type and thickness have to comply with those used in the building.

  Four similar specimens have to be tested. Each specimen has to be linearly supported on four edges (elastomer support, 20 ± 2 mm wide, 10 ± 1 mm thick, Shore A hardness 60 ± 5). The impact body is first dropped from a height of 150 mm onto the centre of the glass specimen. If the specimen does not fail, impact height is gradually increased: 190 mm (Class 3), 450 mm (Class 2), and 1 200 mm (Class 1). The glazing is classified according to the highest drop height at which the test is passed.

  For monolithic glass, the test is successful if the specimen does not break or if it breaks, the glass fragments do not exceed a certain weight. For laminated glass, no openings in the glass larger than 76 mm in diameter are allowed.

- **Germany — TRAV 2003 [327].** As distinct from the other tests described in this paragraph, the experimental setup and the specimen have to be equivalent to the original building unit in terms of materials, support structure etc. Impact testing is done with the standard pendulum according to EN 12600:2002 [105]. Depending on the category of the glazing (cf. Section 2.4), the following drop heights are used: Category A: 450 mm, category B: 700 mm, and category C: 900 mm. The impact body must hit the specimen on points that cause maximum glass and support damage. At least two identical specimens have to be tested.

- **UK — BS 6206 [50].** Impact tests are performed on specimens of 864 mm × 1 930 mm using a 45.36 kg bag filled with lead shot (*Figure 8.1*). The impact classes are defined in function of the drop height. Class C corresponds to a drop height of 305 mm, class B to 457 mm and class A to 1 219 mm.
- **USA — CPSC 16 CFR 1201 [74].** The test setup and testing procedure basically correspond to those in BS 6206. The impact categories (classes), however, are different: Category I corresponds to a drop height of 458 – 470 mm (18 – 18½ in), category II to 1 219 – 1 231 mm (48 – 48½ in). In any case, the glass panels must have a minimum thickness of ¼ in (≈ 6.4 mm).

Figure 8.1:
Impact testing of a broken laminated glass panel shortly after impact (left), approximately 10 min after impact (right).

Examples for overhead glazing

- **Germany — Normal overhead glazing.** There is no specific standard that requires impact testing of overhead glazing, but it is advisable to do such testing if there is a risk of hard objects being dropped on the glass. The impact test is normally performed using a steel sphere. Before testing, the glazing has to be loaded by half the service load, but at least 0.5 kN/m^2. The experimental setup has to be equivalent to the in-service conditions (full scale tests). In Germany, a 4.11 kg steel sphere is dropped from heights between 1 and 3 m. The test is successful if the specimen does not slide from the supports, the impact body does not penetrate the laminated glass and no dangerous glass fragments fall down.
- **Germany — Accessible overhead glazing.** The following guidelines for impact testing of accessible glazing are given in [79]: The experimental set-up has to be equivalent to the in-service conditions (full scale tests). The impact test is performed using a standardized cylindrical steel body weighing 40 kg. The tests are normally done at room temperature, but if very high in-service temperatures are expected, extra tests may be necessary. For the impact test, the specimen is loaded with half of the working load. The load should be applied using 1 kN blocks applied to areas

of 200 mm × 200 mm. This set-up is meant to simulate people standing on the glass surface. The impact body's drop height is 800 mm. It has to hit the specimen on locations that cause maximum glass and support damage. These are usually points of maximum stress and deflection or near supports. The test is successful if the specimen does not slide from the supports, the impact body does not penetrate the laminated glass and no dangerous glass fragments fall down.

- **Germany — Overhead glazing that is accessible for maintenance and cleaning only.** The impact body is a standardized bag filled with glass shot weighing 50 kg [182]. Before testing, the specimen has to be loaded with a single load of 1 kN, applied to an area of 200 mm × 200 mm. This should represent a single person standing on the glass surface. After breakage of the uppermost glass sheet, the whole glazing element must stay on its supports for at least 15 minutes. After this, the impact body must be dropped from a height of 1200 mm to 1800 mm and has to hit the specimen on locations that cause maximum glass and support damage. The criteria to pass the test are identical to those for glazing accessible to the public.

8.1.4 Testing connections

The testing procedure for connections should be chosen as a function of the project-specific needs as well as of the requirements defined by building owners, insurers and authorities. The test regime may include a combination of static and cyclic loads to simulate the intended application. As a minimum the failure load, load history, maximum deformations and the mode of failure should be recorded. As an example, *Figures 8.2* and *8.3* show a bolted connection after failure and the typical failure mechanism.

Figure 8.2:
Example of a bolted connection after failure (laminated glass, 2 × 8 mm, heat strengthened).

8.2 Diagnostic interpretation of glass failures

The failure of architectural glass elements in buildings often impairs the safety and security of a building and its occupants. The failure of glass also has a strong psychological effect on people as broken glass is perceived as a major hazard and such an occurrence triggers a sense of alarm particularly when the cause of failure is not immediately apparent or when the failure seems to be disproportionate to the cause. There is, therefore, a substantial demand for forensic investigations of glass failures. Such failures may be generally classified under one of the following:

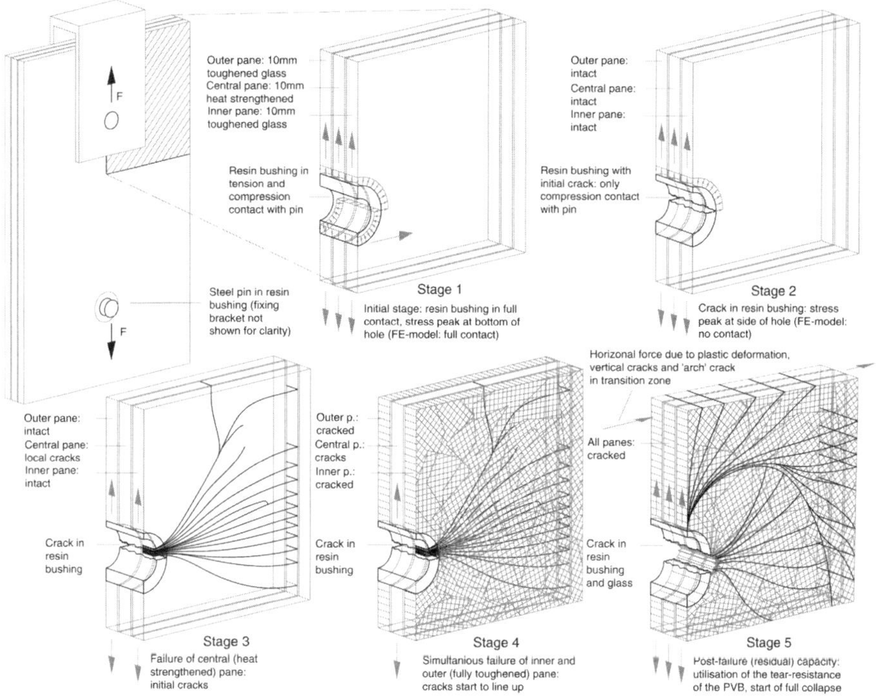

Figure 8.3: Schematic representation of the failure mechanism of a bolted connection in a laboratory test.

- Instability failure, i. e. the glass element lacks adequate lateral fixing or stability or is susceptible to elastic buckling instability such as flexural buckling as encountered in compression members or lateral torsional buckling in the case of flexural members.
- Overstressing of the glass in direct or indirect tension. The overstressing may be caused by excessive uniform loads, blast, impact, thermal stresses or uneven/ inappropriate supports.

It is important to note that any macroscopic flaws or inclusions in the glass will often cause premature failure of the glass at loads that are well within the load bearing capacity expected for a sound glass element. These weaknesses in the glass may either be:

- surface defects (due to macroscopic scratches induced during manufacture or on-site surface damage);
- edge defects (due to poor handling or excessively feathered edges resulting from poor cutting techniques);
- solid inclusions within the thickness of the glass. (This includes nickel sulfide inclusions, which are responsible for spontaneous breakage of tempered glass, however it is important to note that both air bubbles and inclusions other than nickel sulfide often cause failure patterns similar to nickel sulfide failures [193].)

In the event of glass failure, it is often necessary to determine the cause so that liability may be established and to ensure the reliability of the remaining sound glass elements in the building in question and elsewhere. To this end a failure analysis should be undertaken. This typically includes:

1. The collection and review of the history of the use of the glass component (e. g. support conditions, environmental/loading conditions at instant of failure, opportunities for vandalism etc.).
2. A stress analysis model. Often finite element analysis techniques are used (cf. Section 2.3.2).
3. An evaluation of the extent to which the component was used in conformity with specifications.
4. A detailed investigation of the failed glass component (e. g. fractographic and/or chemical analysis).

The first three points above follow standard forensic engineering procedures adopted for most structures and materials and are therefore beyond the scope of this document. Readers should refer to standard texts on this subject, e. g. [10, 261]. On the other hand, the detailed investigation of the failed glass component (i. e. point 4 above) often requires the broad understanding of the factors that influence the fracture patterns and the experience of interpreting these failures.

Fractography, which is the study of fracture surface topography and its relationship to crack propagation, may be very useful in the diagnostic interpretation of glass failure. Fractography techniques normally involve the observation, measurement and interpretation of fracture surfaces in order to determine the origin of failure and the path of the crack thereby giving some insight into the cause of failure. Some of these techniques date back to the observations of *Robert Hooke* who first reported on the fracture surface of limestone in his book 'Micrographia' published in 1665. An excellent review of the wider applications of fractography is given in [199]. Specific fractography applications on glass are given in [45].

8.2.1 Qualitative analysis of failed architectural glass

The first step in the investigation of the failed glass component is the on-site observation and the piecing together of the fragments. This may seem an obvious task, however the broken glass is often disposed of by the building occupants or management and it is important to try and salvage as many of the glass fragments as possible for further analysis. As a minimum it should be possible for the building management to take a picture of the glass before disposing it.

From the failed specimen it is often possible to make some qualitative assessment of the cause of failure by determining the following:

1. The *failure origin*, which helps identify the presence of large flaws or inclusions in the glass, regions of high stress concentration and evidence of bad detailing or possible deliberate damage.
2. The *failure pattern*, which gives an indication of the stresses at failure and the cause of failure. Cracks in annealed glass often nucleate roughly perpendicular to the major principal tensile stresses. The number of flaws or the extent of fragmentation is related to the type of glass used, the surface stress at the instant of fracture, and to the energy imparted to the glass by the action that caused failure (*Figure 8.4*).
3. Specific *topographical features*, which may confirm or dismiss preliminary conclusions reached from the above, e. g. the presence of localized crushing on the surface of the glass close to the failure origin indicates impact from a hard object.

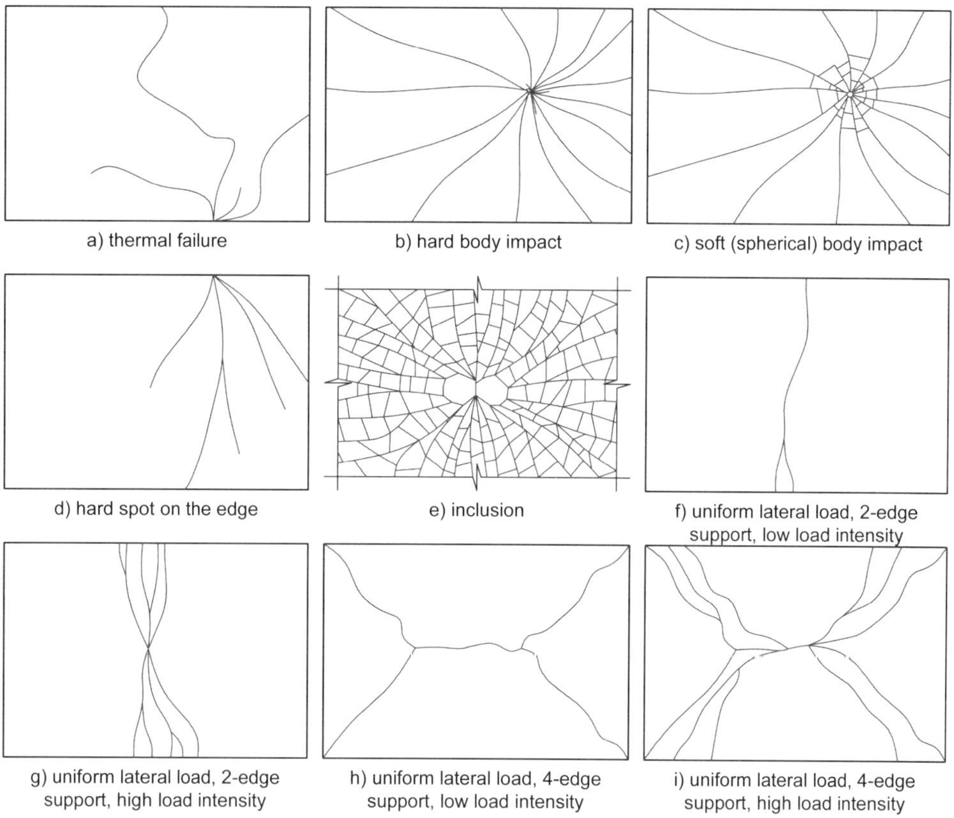

Figure 8.4: Schematic representation of typical glass failures [265].

8.2.2 Quantitative analysis of failed architectural glass

It is desirable to carry out some form of empirical numerical verification of the conclusions drawn from the qualitative analysis of glass failure. From the theoretical review of dynamic fracture presented in Section 3.4, it is possible to obtain an approximation of the surface stress immediately prior to failure. This is done by measuring the crack mirror radius r_{m}, the radius of the mist/hackle boundary r_{h}, or the macroscopic branch length $2r_{\mathrm{b}}$ (see *Figure 3.7*) from the failed glass component and using Equation (3.63) to estimate the corresponding surface stress. From the three determining failure features the crack branching length, $2r_{\mathrm{b}}$, is the simplest one to measure. From the experimental data available (cf. Section 3.4), it may be concluded that a branching constant of $\alpha_{\mathrm{b}} = 2.1\,\mathrm{MPa\,m}^{1/2}$ and an apparent residual stress $\sigma_{\mathrm{ar,b}} = 11\,\mathrm{MPa}$ (annealed glass) would provide good estimates for soda lime silica glass. In the absence of better scientific evidence on how to define the apparent residual stress $\sigma_{\mathrm{ar,b}}$ in heat-treated glass, the actual residual surface compression stress, which is an approximation for $\sigma_{\mathrm{ar,b}}$, should be used. The resulting relationship between failure stress and macroscopic branch length is plotted for all three glass types in *Figure 8.5*. The figure is based on typical residual stress values for heat-treated and fully tempered glass. Since the magnitude of residual stresses can vary considerably, it is advisable to measure the actual residual stress in a broken element of the glass being investigated.

Figure 8.5:
Relationship between failure stress and macroscopic branch length.

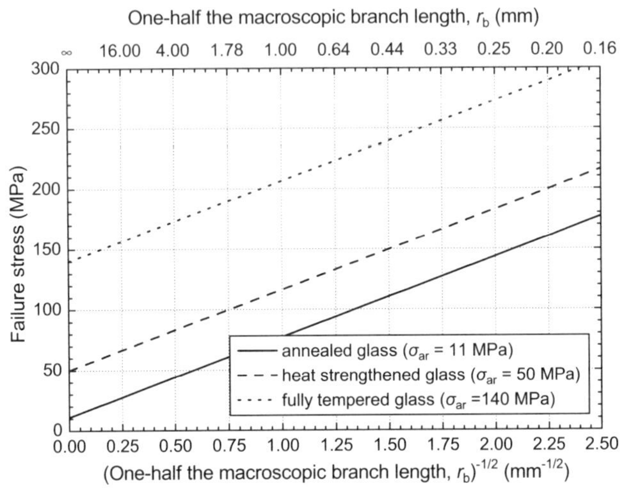

In view of the present scientific evidence, the quantitative relationship between fragment size and fracture stress yields useful results for estimations. However, this should be used with caution as significant gaps in the present knowledge require further research, namely:

- The existing experimental data on heat strengthened and fully tempered glass was obtained on small and thin specimens. Furthermore, the effect of the glass thickness on crack branching requires further investigation.
- Past research focuses on surface flaws. Failures in architectural applications may however be caused by edge flaws. This case needs to be investigated both analytically and experimentally.
- Glass specimens with long surface scratches, such as those found in vandalized glass, may exhibit distorted branching patterns. The macroscopic branch length may be influenced by the scratch and thus produce inaccurate failure stress predictions.

The empirical calculation described above, combined with qualitative observations with the naked eye, is often sufficient to perform a detailed investigation of the failed glass component. In some cases, however, it may be necessary to carry out a second stage of microscopy observations and/or chemical analysis. In glass these observations are carried out by means of optical microscopy or scanning electron microscope (SEM). These additional investigations provide crucial evidence of inclusions in the glass such as solid inclusions and air bubbles. Further investigations such as an energy dispersive X-Ray scan (EDX) will provide an analysis of the chemical composition of the inclusion (e. g. to determine whether it is nickel sulfide or some other form of inclusion). Further details on these techniques are provided in [193, 199].

Appendix

A Notation, Abbreviations

A.1 General information

Variables are defined and explained on their first occurrence only. In case of doubt, readers should refer to the symbol list below. It gives a short description of the variables as well as references to the place where they are defined in the text.

Particularly unfamiliar or important terms are defined in the glossary (p. 183).

The present document follows current regulations on technical and scientific typesetting, in particular [211], [212], [214] and [213]. Accordingly, *italic* symbols are used only to denote those entities that may assume different values. These are typically physical or mathematical variables. Symbols, including subscripts and superscripts, which do not represent physical quantities or mathematical variables are set in upright roman characters. (Example: The exponent '*n*' (italic) in σ_n^n is a physical variable, while the index 'n' (roman) is an abbreviation for 'normal'.)

A.2 Generally used indices and superscripts

$\mathbb{X}_{\mathrm{I,\,II,\,III}}$	related to crack mode I, II or III
$\mathbb{X}_{\mathrm{adm}}$	admissible
$\mathbb{X}_{\mathrm{c}}$	critical
$\mathbb{X}_{\mathrm{d}}$	design level
$\mathbb{X}_{\mathrm{eff}}$	effective
$\mathbb{X}_{\mathrm{eq}}$	equivalent
$\mathbb{X}_{\mathrm{f}}$	failure, at failure, related to failure
$\mathbb{X}_{\mathrm{i}}$	initial
$\mathbb{X}_{\mathrm{inert}}$	in or for inert conditions
$\mathbb{X}_{i}$	*i*-th value, case or time period
$\mathbb{X}_{\mathrm{n}}$	normal, normalized, national
$\mathbb{X}_{\mathrm{test}}$	in laboratory testing, in laboratory conditions
$\sigma^{(i)}$	*i*-th value, case or time period (avoids σ_1 and σ_2, which are the principal stresses)
$\mathbb{X}^{(1)}$	related to a single crack
$\mathbb{X}^{(k)}$	related to *k* cracks

A.3 Functions and mathematical notation

$[\mathbb{X}]$	the unit of $\mathbb{X}$
$\lvert\mathbb{X}\rvert$	the absolute value of $\mathbb{X}$
$\forall$	for all (also ‘for each’ or ‘for every’)
$\exists$	there exists
$\propto$	proportional to
$\gg$	much greater
$\ll$	much less
$\parallel$	parallel to
$f(\mathbb{X})$	a function of the variable $\mathbb{X}$
$\Gamma()$	the Gamma function
max()	maximum
min()	minimum
ln()	natural logarithm
$\mathscr{P}(X)$	probability of the event X ($0 \leq \mathscr{P}(X) \leq 1$)
Φ	the cumulative distribution function of the standard normal distribution

A.4 Abbreviations

4PB	four point bending (test setup)
ANG	annealed glass
BSG	borosilicate glass
CDF	cumulative distribution function
CDR	coaxial double ring (test setup)
FE	finite element
FTG	fully tempered glass
HSG	heat strengthened glass
IPP	in-plane principal stress
LEFM	linear elastic fracture mechanics
PVB	polyvinyl butyral
RSFP	random surface flaw population
SCG	subcritical crack growth
SIF	stress intensity factor
SLSG	soda lime silica glass
SSF	single surface flaw

A.5 Latin symbols

a	1. crack depth → p. 59; 2. longer edge length of a rectangular plate → p. 97
a_0	lower limit of the crack depth → p. 63
a_{i}	initial crack depth → p. 59
a_{c}	critical crack depth → p. 58
a_{f}	crack depth at failure → p. 59
b	short edge length of a rectangular plate → p. 97
c	dimensionless stress distribution function → p. 68
e	eccentricity → p. 110
f_0	reference ambient strength → p. 67
$f_{0,\mathrm{inert}}$	reference inert strength → p. 64
f_{Sd}	design value of the maximum tensile stress → p. 113
f_{Rd}	design value of the maximum tensile strength → p. 113
h	effective glass thickness → p. 97
i, j, k	integer variables

$\bar{k}$	combined parameter (used to simplify notation); → p. 67
$\tilde{k}$	first surface flaw parameter of the glass failure prediction model → p. 96
k_{VSG}	correction factor which takes into account the shear stiffness of the interlayer → p. 124
k_τ	shear buckling coefficient → p. 124
m	number of half sine waves of a buckled plate → p. 123
m_0	second surface condition parameter (see also θ_0) → p. 63
$\bar{m}$	combined parameter (used to simplify notation) → p. 66
$\tilde{m}$	second surface flaw parameter of the glass failure prediction model → p. 96
n	exponential crack velocity parameter → p. 51
q	pressure, uniformly distributed load
$\tilde{q}$	non-dimensionalized load → p. 97
r	parameter of the PDF of the crack depth → p. 63
$\vec{r}$	a point on a surface (defined by two coordinates x and y, $\vec{r} = \vec{r}(x, y)$) → p. 64
t	1. time; 2. glass thickness → p. 109
t_0	reference time period → p. 61
t_{eff}	equivalent glass thickness → p. 111
t_{f}	1. time to failure; 2. point in time when failure occurs; 3. lifetime
t_{int}	interlayer thickness in laminated glass → p. 111
u	displacement → p. 118
v	1. crack velocity → p. 51; 2. lateral deflection of a beam → p. 116
v_0	1. linear crack velocity parameter → p. 51; 2. initial lateral geometric deformation of a beam → p. 109
w	deflection of a bar or a plate → p. 110
w_0	initial deformation of a bar or a plate → p. 109
x, y	coordinates of a point on a surface, cf. $\vec{r}$
z_{a}	distance between the center of gravity and the point where the load is applied → p. 115

A	surface area (general)
A_0	unit surface area ($= 1\,\mathrm{m}^2$) → p. 64
A_{red}	decompressed surface area → p. 88
B	Weibull's risk function → p. 96
E	Young's modulus → p. 56
$EI_{\mathrm{z,eff}}$	equivalent bending stiffness about the z-axis → p. 115
$\hat{F}$	empirical cumulative distribution function → p. 195
G	1. shear modulus → p. 115; 2. energy release rate → p. 56
G_{int}	interlayer shear modulus → p. 111
GK_{eff}	equivalent torsional stiffness → p. 115
I	moment of inertia → p. 110

I_i moment of inertia of layer number i → p. 111

I_S portion of the moment of inertia of a sandwich cross section due to parallel axis theorem → p. 111

$I_\mathrm{S,comp}$ portion of the torsion moment of inertia due to sandwich behaviour → p. 116

I_z moment of inertia about the z-axis → p. 115

K torsion constant → p. 115

K_I stress intensity factor for fracture mode I loading (opening mode) → p. 56

K_Ic fracture toughness (critical stress intensity factor) for fracture mode I loading → p. 57

K_th threshold stress intensity factor → p. 52

L likelihood function → p. 196

L_cr critical buckling length → p. 110

L_LT unrestrained beam length → p. 115

$M_\mathrm{cr,LT}$ critical torsional buckling moment of a beam → p. 115

$M_\mathrm{LT,Rd}$ design value of the bending moment capacity of the glass beam buckling → p. 121

$M_\mathrm{LT,Sd}$ design value of the bending moment due to applied loads → p. 121

N 1. number of samples or other countable quantity; 2. axial compression load → p. 110

N_cr elastic critical buckling load → p. 110

N_Ed design value of applied compression force → p. 128

$N_\mathrm{x,crit}$ critical plate buckling load per unit length → p. 123

P_f failure probability ($= 1 - P_\mathrm{s}$) → p. 63

$P_\mathrm{f,t}$ target failure probability → p. 67

Q force, point load, location- and orientation dependent failure probability

R resistance

S summed square of residuals → p. 197

T duration, time period or end of a time period starting at $t = 0$

U coefficient combining fracture mechanics and crack velocity parameters → p. 66

V_Ed design value of applied compression force → p. 128

W elastic section modulus → p. 110

Y geometry factor (caution: does *not* include $\sqrt{\pi}$; it is $K_\mathrm{I} = Y\sqrt{\pi}\cdot\sigma_\mathrm{n}\cdot\sqrt{a}$) → p. 56

A.6 Greek symbols

α	1. interim parameter to determine the section properties of a sandwich cross section → p. 111; 2. height to width ratio of a plate → p. 123; 3. imperfection factor for buckling curves → p. 129
β	1. shape parameter of the Weibull distribution → p. 76; 2. interim parameter to determine the section properties of a sandwich cross section → p. 111
γ	1. partial factor (specified more precisely in the index); 2. shear deformation → p. 156
θ	general Weibull scale parameter (specified more precisely in the index)
θ_0	first surface condition parameter (see also m_0) → p. 63
$\overline{\lambda}_{\mathrm{LT}}$	slenderness ratio of a beam → p. 120
$\overline{\lambda}_{\mathrm{P}}$	slenderness ratio of a beam → p. 127
μ	mean
ν	Poisson's ratio → p. 56
ρ	reduction factor for plate buckling factor → p. 127
σ	1. stress (details specified in the index); unless otherwise stated, compressive stresses are negative and tensile stresses positive; 2. standard deviation
$\dot{\sigma}$	stress rate ($\dot{\sigma} = \mathrm{d}\sigma/\mathrm{d}t$)
$\bar{\sigma}$	equivalent reference stress → p. 67
$\breve{\sigma}$	representative stress (often $\sigma_{\max}$) → p. 68
$\tilde{\sigma}$	non-dimensionalized stress → p. 97
σ_1	major in-plane principal stress ($\sigma_1 \geq \sigma_2$) → p. 64
σ_2	minor in-plane principal stress ($\sigma_1 \geq \sigma_2$) → p. 64
σ_{c}	inert strength of a crack (also called 'critical stress') → p. 58
σ_{t_0}	t_0-equivalent static stress → p. 61
$\sigma_{\mathrm{cr,LT}}$	critical lateral torsional buckling stress → p. 120
σ_E	surface stress due to action(s) E → p. 57
σ_{f}	stress at failure (also known as 'failure stress' or 'breakage stress')
$\sigma_{\max}$	maximum principal stress in an element (geometric maximum) → p. 68
σ_{n}	in-plane surface stress normal to a crack's plane (also known as the crack opening stress) → p. 64
σ_{r}	residual surface stress due to tempering (sometimes called 'prestress'; compressive ⇒ negative sign) → p. 57
σ_{R,t_0}	t_0-equivalent resistance → p. 61
σ_{Rk}	characteristic tensile strength → p. 120
σ_{p}	1. stress due to external constraints or prestressing → p. 57; 2. compressive edge stress of a plate subjected to buckling → p. 128
φ	crack orientation, angle → p. 64
τ	1. time (point in time); 2. shear stress (general)
τ_{crit}	critical buckling load of a plate subjected to shear → p. 124
ψ	load combination factor → p. 100
χ_{LT}	reduction factor for lateral torsional buckling → p. 121

Appendix

B Glossary of Terms

Action General term for all mechanical, physical, chemical and biological actions on a structure or a structural element, e. g. pressures, loads, forces, imposed displacements, constraints, temperature, humidity, chemical substances, bacteria and insects.

Action history The description of an action as a function of time.

Abhesive A material that resists adhesion; a film of coating applied to surfaces to prevent sticking, heat sealing, and so on, such as a parting agent or mold release agent.

Abrasion (general) The wearing away of a material surface by friction.

Abrasion (decorative glass) A method of shallow, decoration grinding using a diamond wheel.

Absolute humidity The weight of water vapour present in a unit of air.

Accelerated ageing Any set of test conditions designed to determine, in a short time, the result obtained under normal conditions of ageing. In accelerated ageing tests, the usual factors considered are heat, light, and oxygen, either separately or combined.

Accelerated weathering Machine-made means of duplicating or reproducing weather conditions. Such tests are particularly useful in comparing a series of products at the same time. No real correlation between test data and actual service is known for resins and rubbers used in many products.

Acid etching A process, manly used for glass decoration, where the glass surface is treated with hydrofluoric acid. Acid-etched glass has a distinctive, uniformly smooth and satin-like appearance.

Acoustical double glazing Two monolithic glass panels, set in a frame, with an air space between them.

Acrylate resins Polymerization products of certain esters of acrylic and methacrylic acid, such as methyl or ethyl acrylate. Possess great optical clarity and high degree of light transmission. Nearest approach to an organic glass.

Acrylic A group of thermoplastic resins or polymers formed by polymerizing the esters of acrylic acid.

Action intensity The magnitude of an action, e. g. a load intensity, a stress intensity or the magnitude of an imposed deformation. See also 'load shape'.

Active solar heat gain Solar heat that passes through a material and is captured by mechanical means.

Adduct A chemical addition product.

Adhere That property of a sealant/compound which measures its ability to bond to the surface to which it is applied.

Adhesion The clinging or sticking of two material surfaces to each other. In rubber parlance, the strength of the bond or union between two rubber surfaces or plies, cured or uncured. The bond between a cured rubber surface and non-rubber surface, e.g., glass, metal, wood, or fabric.

Adhesion failure (1) The separation of the two surfaces with a force less than specified. (2) The separation of the two adjoining surfaces due to service conditions.

Adhesive setting Classifies the conditions to convert the adhesive from its packaged state to a more useful form.

Adsorption The action of a body in condensing and holding gases, dyes, or other substances. The action is usually considered to take place only at or near the surface. The power of adsorption is one of the characteristic properties of matter in the colloidal state and is associated with surface energy phenomena of colloidally dispersed particles.

Ageing A progressive change in the chemical and physical properties of rubber, especially vulcanized rubber, usually marked by deterioration. The verb is also used transitively to denote the setting aside of rubber goods under specified conditions for the purpose of observing their rate of deterioration.

Ageing resistance Resistance to ageing by oxygen and ozone in the air, by heat, and by light.

Ageing tests Accelerated tests of rubber specimens to find out their endurance by heating them in air under pressure or similarly in oxygen.

Air infiltration The amount of air that passes between a window sash and frame or a door panel and frame.

Air side In the float process, the upper side of glass is called the air side.

Alkali Substance that neutralizes acid to form salt and water. Yields hydroxyl (OH-) ions in water solution. Proton acceptor.

Ambient noise The all-encompassing noise associated with a given environment, usually a composite of sounds from sources near and far.

Ambient temperature The environmental temperature surrounding the object.

Annealing The process which prevents glass from shattering after it has been formed. The outer surfaces of the glass shrink faster than the glass between the surfaces, causing residual stresses which can lead to shattering. This can be avoided by reheating the glass and allowing it to cool slowly.

Artificially induced surface damage Any kind of damage that is induced systematically and on purpose, e. g. for laboratory testing. If it is homogeneous in terms of its characteristics and its distribution on the surface, it is called 'artificially induced homogeneous surface damage' (e. g. surface damage induced by sandblasting).

Antiwalk blocks Rubber blocks that prevent glass from moving sideways in the glazing rabbet because of thermal effects or vibration.

Art glass Art glass goes by many names. It is called opalescent glass, cathedral glass, or stained glass and is usually produced in small batch operations.

Artificially induced damage Any kind of damage that is induced systematically and on purpose, e. g. for testing purposes.

As-received glass Glass as it is delivered to the client, sometimes also called 'new glass'. The surface contains only the small and random flaws introduced by production, cutting, handling and shipping.

Aspect ratio The relationship between the long and the short edge lengths of a rectangular plate.

Attenuation The reduction of sound pressure level, usually expressed in decibels.

Autoclave A vessel that employs heat and pressure. In the glass industry, used to produce a bond between glass and PVB or urethane sheet, thus creating a laminated sheet product.

Back-fill Placing material into the opening between glass and glazing.

Bait A webbed metal frame used to draw molten glass.

Bandage joint Sealant joint composed of bond-breaker tape over the joint movement area with an overlay of sealant lapping either side of the tape sufficient to bond well to the surfaces; often used where extreme movement occurs and conventional joint design is not possible.

Batch The mixed raw materials which are used to make glass.

Bead A sealant/compound after application in a joint. Also a molding or stop used to hold the glass product in position.

Bent glass Flat glass shaped while hot into cylindrical or other curved shapes.

Bevel or compound bead In glazing, a bead of compound applied to provide a slanted top surface so that water will drain away from the glass or panel.

Bevelling The process of edge-finishing flat glass by forming a sloping angle to eliminate right-angled edges.

Bifurcation buckling If the load applied on a structural member exceeds the critical value, the straight position is unstable and a slight disturbance leads to large displacements and, finally, to the collapse of the member by buckling. The critical point, after which the deflections of the member become very large, is called the "bifurcation point" of the system.

Bite The dimension by which the edge of a glass product is engaged into the glazing channel.

Body-tinted glass See tinted glass.

Blocks Rectangular, cured sections of neoprene or other approved materials, used to position a glass product in a glazing channel.

Bond (noun) (1) The attachment at the interface between an adhesive and an adherent. (2) A coat of finishing material used to improve the adhesion of succeeding coats.

Bond (verb) To join materials together with adhesives. To adhere.

Bond breaker A material to prevent adhesion at a designated interface.

Bond strength The force per unit area or length necessary to rupture a bond.

Bonding agents Substances or mixtures of substances used in attaching rubber to metal.

Bow A continuous curve of a glass sheet, either vertical or horizontal.

Breather tube A small-diameter tube placed into the space of an insulating glass unit through the perimeter wall to equalize the air pressure within the unit. These tubes are to be sealed on the job site prior to installation.

Bronze glass A glare- or heat-reducing glass intended for applications where glare control and reduction of solar heat are desired or where colour can contribute to design.

Buckling Buckling is a failure mode characterized by a sudden failure of a structural member that is subjected to high compressive stresses where the actual compressive stresses at failure are smaller than the ultimate compressive stresses that the material is capable of withstanding. This mode of failure is also described as failure due to elastic instability.

Buckling curve Buckling curves afford a means of design aid for stability critical structural elements taking into account geometrical, structural and material imperfections.

Bullet-resistant glazing Security glazing affording a defined resistance against the firing of specified weapons and ammunition.

Bull's eye The round, whorl shape in the center of old panes of glass.

Butt glazing The installation of glass products where the vertical glass edges are without structural supporting mullions.

Butt joint A joint having opposing faces which may move toward or away from one another; a joint in which the receiving surfaces stresses the sealant in tension or compression.

Butyl rubber A copolymer of about 98% isobutylene and 2% isoprene. It has the poorest resistance to petroleum oils and gasolines of any rubber. Excellent resistance to vegetable and mineral oils, to solvents, such as acetone, alcohol, phenol, and ethylene glycol, and to water and gas absorption. Heat resistance is above average. Sunlight resistance is excellent. Its abrasion resistance is not as good a natural rubber. Usually low permeability to gases.

Chain polymerization A chain reaction in which the growth of a polymer chain proceeds exclusively by reaction(s) between monomer(s) and reactive site(s) on the polymer chain with regeneration of the reactive site(s) at the end of each growth step.

Channel A three-sided U-shaped opening in sash or frame to receive pane or panel, with or without removable stop or stops. Contrasted to a rabbet, which is a two-sided, L-shaped section, as with face-glazed window sash.

Channel glazing The sealing of the joints around panes or panels set in a U-shaped channel employing removable stops.

Chemically strengthened glass Glass with a residual compressive surface stress produced by a process of ion exchange.

Chemical resistance The resistance offered by elastomer products to physical or chemical reactions from contact with or immersion in various solvents, acids, alkalies, salts, etc.

Clips Wire spring devices to hold glass in a rabbet sash, without stops or face glazing.

Coating A material, usually liquid, used to form a covering film over a surface. Its function is to decorate, to protect the surface from destructive agents or environments (abrasion, chemical action, solvents, corrosion, and weathering) and/or to enhance the (optical, mechanical, thermal) performance.

Coefficient of variation (CoV) A measure of dispersion of a probability distribution. It is defined as the ratio of the standard deviation σ to the mean μ.

Coefficient of expansion The coefficient of linear expansion is the ratio of the change in volume per degree to the length at 0 °C. The coefficient of volume expansion (for solids) is three times the linear coefficient. The coefficient of volume expansion for liquids is the ratio of the change in volume per degree to the volume at 0 °C.

Cohesive failure The splitting and opening of a sealant/compound within its body, resulting in water penetration.

Cold resistant Withstands the effect of cold or low temperatures without loss of serviceability.

Colour cast glass Includes many kinds of cast and rolled glass. There are more than 100 colours.

Column buckling Column buckling is defined as an instance of lateral bending of a bar due to a axial compressive load.

Computing time The time required to run an algorithm on a computer. While the actual value depends on the performance on the computer, the term is still useful for qualitative considerations and comparisons.

Condensation Moisture that forms on surfaces colder than the dew point.

Conduction The transfer of heat through matter, whether solid, liquid, or gas.

Consistency The viscosity or solidness of a semisolid or syrupy substance. It may be called the resistance to deformation. That property of a body by which it resists deformation or permanent change of shape.

Constant stress rate loading A specimen is loaded such that the stress increases linearly with respect to time.

Constant load rate loading The load on a specimen is increased linearly with respect to time.

Convection A transfer of heat through a liquid or gas, when that medium hits against a solid.

Crack In the present document, the term 'crack' refers to the idealized model of a flaw having a defined geometry and lying in a plane.

Chromogenics Any visibly switchable technology useful for glazing, mirrors and transparent displays.

Cullet Recycled or waste glass.

Cure To change the properties of a material by chemical reaction, which may be condensation, polymerization, or vulcanization. Usually accomplished by the action of heat and catalysts, alone or in combination, with or without pressure.

Curtain walling Non-load bearing, typically aluminium, façade cladding system, forming an integral part of a building's envelope.

Curved glass Glass, which is curved in form, produced by heating it to its softening point, so that it takes the shape of the mould.

Damping The dissipation of sound energy in a medium over time or distance.

Decompressed surface The part of an element's surface where the tensile stress due to loading is greater that the residual compressive stress due to tempering. On these parts of the surface, there is a positive crack opening stress.

Defect A flaw that is unacceptable.

Deflection The physical displacement of glass from its original position under load.

Deformation Any change of form or shape produced in a body by stress or force.

Degradation Deterioration, usually in the sense of a physical or chemical process, rather than a mechanical one.

Dehydration Removal of water as such from a substance, or after formation from a hydrogen and hydroxyl group in a compound, by heat or dehydrating substance.

Delaminate To split a laminated material parallel to the plane of its layers. Sometimes used to describe cohesive failure of an adherent in bond strength testing.

Delamination Separation or splitting, usually lack of adhesion in plied goods.

Desiccants Porous crystalline substances used to absorb moisture and solvent vapours from the air space of insulating glass units. More properly called absorbents.

Design See structural design.

Design life The period of time during which a structural element is expected to perform according to its specification, i. e. to meet the performance requirements.

Dew point The temperature at which air is saturated with respect to a condensible component, such as water vapour or solvent.

Discoloration Staining. Changing or darkening in colour from the standard or original.

Double glazing, double-glazed units See insulating glass unit.

Drawing tower Used in the sheet glass process for drawing molten glass.

Dual sealed system A primary seal of polyisobutylene and a secondary seal of polysulphide, polyurethane or silicone ensure the effective and durable seal of double-glazed units.

Edge clearance The distance between the edge of the glass and rebate.

Edge joint A joint made by bonding the edge faces of two adherends.

Effective nominal flaw depth The depth of a flaw that is calculated from its measured strength.

Elasticity The property of matter which causes it to return to its original shape after deformation, such as stretching, compression, or torsion.

Elastomer A substance that can be stretched to at least twice its original length and, after having been stretched and the stress removed, returns to approximately its original length in a short time.

Elongation Increase in length expressed numerically as a fraction or percentage of initial length.

Emissivity The relative ability of a surface to absorb and emit energy in the form of radiation. Emissivity factors range from 0.0 (0%) to 1.0 (100%).

Emittance Heat energy radiated by the surface of a body, usually measured per second per unit area.

Enamel A soft glass compound of flint or sand, soda potash, and red lead. It is the colourful result of fusion of powdered glass to a substrate through the process of firing, usually between 750 °C and 850 °C. The powder melts and flows to harden as a smooth, durable vitreous coating on metal, glass or ceramic.

Enamelled glass Enamelled glass is tempered or heat strengthened glass, one face of which is covered, either partially or totally, with mineral pigments.

Energy absorptance The percentage of solar radiant heat energy absorbed and re-emitted externally and internally by the glass.

Energy reflectance (RE) The percentage of solar radiant heat energy reflected by glazing.

EPDM EPDM rubber (ethylene propylene diene monomer rubber) is an elastomer which is characterized by wide range of applications (i. e. as automotive weather-stripping and seals, glass-run channel, garden and appliance hose, tubing, washers, roofing membrane, geomembranes, rubber mechanical goods).

Equibiaxial stress field The two in-plane principal stresses are equal ($\sigma_1 = \sigma_2$). In this stress state, the stress normal to a crack σ_n is independent of the crack's orientation φ_{crack}, meaning that $\sigma_n = \sigma_1 = \sigma_2 \ \forall \varphi_{crack}$. An equibiaxial stress field is in particular found within the loading ring in coaxial double ring testing.

Exterior glazed Glass set from the exterior of the building.

Exterior stop The removable molding or bead that holds the pane or panel in place when it is on the exterior side of the pane or panel, as contrasted with an interior stop located on the interior side of the pane.

Extruded Forced through a die or continuous mold for shaping.

Face Describes the surfaces of the glass in numerical order from the exterior to the interior. The exterior surface is always referred to as face 1. For a double-glazed unit, the surface of the outer pane facing into the cavity is face 2, the surface of the inner pane facing into the cavity is face 3 and the internal surface of the inner pane is face 4.

Flat glass Pertains to all glass produced in a flat form.

Flaw General term describing a condition or change that indicates an abnormal condition or imperfection in a material. Only flaws that are unacceptable are defects.

Float glass Transparent glass with flat, parallel surfaces formed on the surface of a bath of molten tin. If no information with respect to heat treatment is given, the term generally refers to *annealed* float glass.

Fogged unit An insulating glass unit with a permanent deposit that contaminates its interior surfaces.

Forming Shaping or molding into shape.

Front putty The putty forming a triangular fillet between the surface of glass and the front edge of the rabbet.

Frost point The temperature below 0 °C at which visible frost begins to deposit on the air-space surface of a sealed insulating glass unit.

Frosted glass Glass produced by acid etching or sand blasting. These surface modifications have the effect of rendering the glass translucent, obscuring the view while still passing light.

Fully tempered glass Glass with a high residual compressive surface stress, varying typically between 80 MPa and 150 MPa in the case of soda lime silica glass. According to ASTM C 1048-04 [12], fully tempered glass is required to have either a minimum surface compression of 69 MPa (10 000 psi) or an edge compression of not less than 67 MPa (9 700 psi). In European standards, the fragmentation count and the maximum fragment size is specified [101, 102].

g Abbreviation or symbol for 'solar factor' according to EN 410:1998 [147].

Gas-filled units Insulating glass units with a gas other than air in the air space to decrease the unit's thermal conductivity (U-value) and to increase the unit's sound insulating value.

Gasket Pre formed shape, such as a strip, grommet, etc., of rubber and rubber-like composition used to fill and seal a joint or opening, alone or in conjunction with the supplemental application of a sealant.

Glass A uniform amorphous solid material, usually produced when a suitably viscous molten material cools very rapidly to below its glass transition temperature, thereby not giving enough time for a regular crystal lattice to form. By far the most familiar form of glass is soda lime silica glass. In its pure form, glass is a transparent, relatively strong, hard-wearing, essentially inert, and biologically inactive material which can be formed with very smooth and impervious surfaces. Glass is, however, brittle and will break into sharp shards. These properties can be modified, or even changed entirely, through the addition of other compounds or heat treatment.

Glazing The securing of glass into prepared openings. It also refers to the collective elements of a building comprising glass, frame and fixings.

Glazing bead A strip surrounding the edge of the glass in a window or door; applied to the sash on the outside, it holds the glass in place.

Glazing channel A three-sided U-shaped sash detail into which a glass product is installed and retained by a removable stop.

Glue Historically, glue only refers to protein colloids prepared from animal tissues. The meaning has been extended to any type of glue-like substances that are used to attach one material to another.

Greenhouse glass This is a translucent rolled glass with a special surface design to scatter light.

Guarding The prevention of people falling wherever there is a change in floor level by means of a permanent barrier.

Hardness Property or extent of being hard. Measured by extent of failure of indentor point of any one of a number of standard testing instruments to penetrate the product.

Heat-absorbing glass Glass (usually tinted) formulated to absorb an appreciable portion of solar energy.

Heat-soak test (HST) A heat-treatment which is carried out after the tempering process in order to reduce the risk of spontaneous breakage of heat-treated glass in service due to nickel sulfide inclusions.

Heat strengthened glass Glass with a medium residual compressive surface stress. Heat strengthened glass is required, according to [12], to have a residual compressive surface stress between 24 MPa (3 500 psi) and 52 MPa (7 500 psi). In European standards, the fragmentation count and the maximum fragment size is specified [133, 134].

Heat-treated glass Glass that has been thermally treated to some extent. The term includes heat strengthened and fully tempered glass.

Heel bead Sealant applied at the base of channel, after setting pane or panel and before the removable stop is installed.

Homogeneous The opposite of heterogeneous. Consisting of the same element, ingredient, component, or phase throughout, or of uniform composition throughout.

Immersion Placing an article into a fluid, generally so it is completely covered.

Impact The single instantaneous stroke or contact of a moving body with another either moving or at rest.

Impact strength Measure of toughness of a material, as the energy required to break a specimen in one blow.

Infill panel The glass panel underneath the handrail in a barrier that provides containment, but no structural support to the main frame of the barrier.

Inherent strength The part of the tensile strength that is *not* due to compressive residual stresses but to the resistance of the material itself. For float glass, this is approximately (even float glass has some compressive residual stresses) the measured macroscopic resistance.

Inner pane The pane of a double-glazed unit which faces the interior of a building.

Insulating glass unit (IGU) A piece of glazing consisting of two or more layers of glazing separated by a spacer along the edge and sealed to create a dead air space or a vacuum between the layers in order to provide thermal insulation. The dead air space is often filled with inert gas (argon or, less commonly, krypton).

Intaglio A light engraving on the surface of glass.

Integrity The ability of glazing to remain complete and to continue to provide an effective barrier (e. g. to flames or people).

Interior glazed Glass set from the interior of the building.

Interior muntins Decorative grid installed between the glass panes that does not actually divide the glass.

Interior stop The removable molding or bead that holds the pane in place, when it is on the interior side of the pane.

Interlayer Very thin layer between two materials. In laminated glass: a transparent, tough plastic sheeting material, such as PVB, that is able to retain the fragments after fracture.

Intumescence The swelling and charring of materials when exposed to fire.

Joint The location at which two adherents are held together by an adhesive.

Laminated glass Two or more panes of glass bonded together with a plastic interlayer.

Lap joint A joint made by overlapping adjacent edge areas of two adherents to provide facing surfaces which can be joined with an adhesive.

Lateral load Short form of 'out-of-plane load', often also used as a short form for → uniform lateral load.

Lateral torsional buckling Lateral torsional buckling is defined as an instance of lateral bending about the weak cross section axis of a bar due to bending about the strong axis.

Lehr Similar to an oven, used to anneal glass by reheating it and allowing it to cool slowly.

Light reducing glass Glass formulated to reduce the transmission of visible light.

Light reflectance The proportion of the visible spectrum that is reflected by the glass.

Light transmittance The proportion of the visible spectrum that is transmitted through the glass. Clear glass, depending on its thickness, allows 75 to 92% of visible light to pass through.

Lite Another term for a pane of glass.

Load duration factor The effect of a given load depends not only on its intensity, but also on the duration of a glass element's exposure to the load. This is often accounted for by applying a duration-dependent factor, the so called 'load duration factor', either to the load intensity or to some reference resistance.

Load shape Describes the geometric properties of a load, e. g. whether it is a distributed load, a point load, a line load, or a free-form load, where on a structural element it is applied and whether it is uniform, triangular or has some other shape. A complete characterization of a load must include its shape and intensity (cf. 'action intensity').

Loading time The time period during which a load is applied.

Low emissivity coating (low-e coating) A transparent metallic or metallic oxide coating that saves energy and increases comfort inside a building by reducing heat loss to the environment.

Low iron glass Extra clear glass, which has a reduced iron oxide content in order to lessen the green tinge inherent in ordinary clear float glass.

Metal spacers Roll-formed metal shapes used at the edges of an insulating glass unit to provide the designated air-space thickness.

Mode I Loading condition that displaces the crack faces in a direction normal to the crack plane, also known as the opening mode of deformation.

Monotonously increasing If $x(t)$ is monotonously increasing with t, it is $x(t_2) > x(t_1)$ for any $t_2 > t_1$. The increase may or may not be linear.

Mullion A horizontal or vertical member that holds together two adjacent panes of glass or units of sash or sections of curtain wall.

Multiple-glazed units Units of three panes (triple-glazed) or four panes (quadruple-glazed) with two and three dead air spaces, respectively.

Muntin In sash having horizontal and vertical bars that divide the window into smaller panes of glass, the bars are termed muntin bars. Similar to mullion but lighter in weight.

Neoprene A synthetic rubber with physical properties closely resembling those of natural rubber but not requiring sulfur for vulcanization.

Nickel sulfide inclusion A rare, but naturally occurring impurity present in all glass that can, in certain circumstances, lead to spontaneous breakage of heat-treated glass in service.

Non-uniform stress field A stress field in which the stress varies from one point of the surface to another (cf. uniform stress field).

Off-line coating See sputtered coating.

On-line coating See pyrolytic coating.

Opaque glass Glass that transmits no light whatsoever.

Opaline glass This glass is closely related to opaque. It is an opaque cast with ground and polished surfaces.

Ornamental glass Rolled glass with the surface figured by shaping or embossing rolls.

Outer pane The pane of a double-glazed unit which faces the exterior of a building.

Pane (of glass) A sheet of glass.

PEEK Polyether ether ketone (PEEK) is a partially crystalline polymer. It has repeating monomers of two ether and ketone groups and is one of the highest rated thermoplastic materials in terms of heat resistance. PEEK has excellent chemical resistance, high strength and good resistance to burning. Its disadvantage is the high cost.

Predictive modelling The creation of a new model or the use of an existing model to predict the behaviour of a system, e. g. the mechanical behaviour of a structural glass element.

Passive solar heat gain Solar heat that passes through a material and is captured naturally, not by mechanical means.

Patterned glass Rolled glass with an embossed pattern on one or both surfaces.

Peeling The loosening of a rubber coating or layer from a base material, such as cloth or metal, or from another layer of rubber.

Permanent set The amount by which an elastomeric material fails to return to its original form after a deformation.

Permeability The degree of water vapour or gas transmission through a unit area of material of unit thickness induced by unit vapour pressure differences between two specific surfaces under specified temperature and humidity conditions.

Permeance The time rate of water vapour or gas transmission through a unit area of a body, normal to specific parallel surfaces, under specific temperature and humidity conditions.

Plastics Natural and artificially prepared organic polymers of low extensibility, as compared with rubber, which can be molded, extruded, cut, and worked into a great variety of objects, rigid or nonrigid, and used as substitutes for wood, metals, glass, rubber, leather, fibers, and textile materials. Many are also referred to as synthetic resins.

Plate buckling Plate buckling is defined as an instance of out of plane bending of a plate due to in plane compressive stress.

Points Thin, flat, triangular, or diamond-shaped pieces of zinc used to hold glass in wood sash by driving them into the wood.

POM Polyoxymethylene (POM), also known as acetal resin, polytrioxane, polyformaldehyde, and paraformaldehyde, is an engineering plastic used to make gears, bushings and other mechanical parts.

Potash Potassium oxide (a flux).

Primary seal A butyl-based sealant, for example polyisobutylene, applied to the edges of the spacer bar during assembly into double-glazed units, to ensure a watertight and airtight seal around the perimeter of the unit.

Primer A special coating designed to provide adequate adhesion of a coating system to a new surface.

Priming Sealing of surfaces to produce adhesion of sealants.

Profile glass Usually U-shaped, rolled glass for architectural use.

Purlins Structural members, generally horizontal, on sloped glazing frames.

PVB (polyvinyl butyral) Polyvinyl butyral is a viscoelastic resin that is made from vinyl acetate monomer as the main raw material. It provides strong binding, optical clarity, adhesion to many surfaces, toughness and flexibility. PVB is the most commonly used interlayer material for laminated glass.

Pyrolytic coating A metallic coating applied to the glass 'on-line' during the float glass manufacturing process. The high temperatures involved result in the metallic oxides fusing into the surface of the glass through pyrolysis.

R-value The resistance of conductive heat energy transfer of a specific insulating glass unit assembly. It is the reciprocal of the U-value ($R = 1/U$).

Rafters Structural members; vertical in sloped glazing frames.

Radiation Energy released in the form of waves or particles because of a change in temperature within a gas or vacuum.

Rebate The section of the frame surround which forms an angle into which the glass is placed and held.

Reflective coating A metallic or metallic oxide coating applied to one side of the glass in order to significantly increase the amount of reflection of both the visible and infrared range of the electromagnetic spectrum.

Relative heat gain An energy comparison factor for glass products combining the radiant and conductive heat gain under specific conditions.

Residual stress The residual compressive surface stress that arises from the tempering process. (The term 'pre-stress' is, although widely used, somewhat misleading and therefore not used in the present document.)

Resin laminate Two or more sheets of glass assembled with one or more resin interlayers.

Rheology Science of deformation and flow of matter. Deals with laws of plasticity, elasticity, viscosity, and their connection with paints, plastics rubber, oils, glass, cement, etc.

Rigidity The property of bodies by which they can resist an instantaneous change of shape. The reciprocal of elasticity.

Rollerwave An optical phenomenon, generally noticed in reflection, caused by contact between glass and rollers in the horizontal tempering process.

Sandblasting A special glass treatment in which sand is sprayed at high velocities over the surface of the glass.

Sash A frame into which glass products are glazed, i. e., the operating sash of a window.

Score side The upper side of glass coming off the float line, sometimes called the air side.

Screen printed glass Tempered or heat strengthened glass, one face of which is covered, either partially or completely, with a mineral colour.

Sealant A material used to fill a joint, usually for the purpose of weather-proofing or waterproofing. It forms a seal to prevent gas and liquid entry.

Sealants (for insulating glass units) Formulated elastomeric compounds with specific application and vapour transmission properties as well as controlled adhesion, cohesion, and resiliency.

Secondary seal A sealant, usually polysulphide, polyurethane or silicone, applied to the edges of double-glazed units after the primary seal, to provide effective and durable adhesion between the glass components and spacer bar.

Setting Placement of panes or panels in sash or frames.

Setting blocks Small blocks of composition, lead, neoprene, wood, etc., placed under the bottom edge of the pane or panel to prevent its settling onto the bottom rabbet or channel after setting, thus distorting the sealant.

Shading coefficient The solar factor (total transmittance) of a glass relative to that of 3 mm clear float glass. Used as a performance comparison. The lower the shading coefficient, the lower the amount of solar heat transmitted. The short wave shading coefficient is the direct transmittance (T) of the glass as a factor of the solar factor or total transmittance (g or TT) of 3 mm clear float glass. The long wave shading coefficient is the internally re-radiated energy that the glass has absorbed as a factor of the solar factor (total transmittance) of 3 mm clear float glass. It is determined by subtracting the direct transmittance from the solar factor (total transmittance) of the subject glass and then dividing by the solar factor (total transmittance) of 3 mm clear float glass.

Silica Silica, also known as silicon dioxide (SiO_2), is a very common mineral composed of silicon and oxygen. Quartz and opal are two forms of silica. In nature, silica is commonly found in sand.

Silicates Silicates are minerals composed of silicon and oxygen with one or more other elements. Silicates make up about 95% of the Earth's crust.

Silicone seal Where the edges of double-glazed units are unframed and exposed to direct sunlight, they are sealed with silicone for UV resistance.

Silvering A process used in the manufacture of mirrors, whereby a silver coating is applied to one surface of the glass.

Skylight A glass and frame assembly installed into the roof of a building.

Slenderness ratio The slenderness ratio is a means of classifying structural members (columns, beams, plates) with respect to their risk of failure due to instability.

Sloped glazing Any installation of glass that is at a slope of 10°/ 15° (depends on the standard) or more from the vertical.

Solar control coating A coating that absorbs or reflects solar energy.

Solar energy absorption The percentage of the solar spectrum energy that is absorbed by a glass product.

Solar factor g The percentage of total solar radiant heat energy transmitted through glazing (the sum of energy transmitted directly and energy absorbed and re-emitted to the interior).

Solar heat gain Solar radiant heat, transmitted or re-emitted by glazing into a building, contributing to the build-up of heat.

Sound reduction index A laboratory measure of the sound insulating properties of a material or building element in a stated frequency band.

Spacer, spacer bar Generally an aluminium bar along all edges of a double-glazed unit, filled with desiccant, which separates the two panes of glass and creates a cavity.

Spacers Small blocks of composition, neoprene, etc., placed on each face of pane and panel to center them in the channel and maintain uniform width of sealant beads, preventing excessive sealant distortion.

Spall Small fragments of glass that are ejected from the surface of a laminated glass sheet when the opposite surface is impacted.

Spandrel, spandrel panel Glass cladding panels used in non-vision areas of a façade, commonly in curtain walling. They generally comprise an enamelled or opacified glass to conceal building structure elements such as the edge of floor slabs.

Sputtered coating A coating applied to the glass 'off-line' or after the float glass manufacturing process by a technique called magnetron sputtering under vacuum conditions.

SSG Structural sealant glazing.

SSGS Structural sealant glazing systems.

Stability Stability theories are formulated in order to determine the conditions under which a structural system, which is in equilibrium, ceases to be stable. Instability is essentially a property of structures in their extremes of geometry; for example, long slender struts, beams, thin flat plates or thin cylindrical shells. In very general terms, stability may be defined as the ability of a physical system to return to equilibrium when slightly disturbed.

Starved joint A joint that has an insufficient amount of adhesive to produce a satisfactory bond.

Stepped-edge unit The edges of the double-glazed unit are not flush. One pane is larger and overlaps the other, to enable their use in roof glazing for example.

Stop Either the stationary lip at the back of a rabbet or the removable molding at the front of the rabbet serving to hold the pane or panel in sash or frame with the help of spacers.

Strength The maximum stress required to overcome the cohesion of a material. Strength is considered in terms of compressive strength, tensile strength, and shear strength, namely the limit states of compressive stress, tensile stress and shear stress respectively.

Stress rate The stress rate $\dot{\sigma}$ is the increase in stress per unit of time, or, in other words, the derivative over time of the stress: $\dot{\sigma} = \mathrm{d}\sigma/\mathrm{d}t$.

Structural design The iterative process of selecting a structural element that meets a set of performance requirements that depend on the specific application. Common requirements for structural glass elements relate to aspects such as deformation, vibration, usability, aesthetics, acoustic or optical performance, and, of course, load bearing capacity.

Structural glazing Glass acting as a structural support to other parts of the building structure. It can also refer to glass that is fixed by means of bolted connectors, although the glass is not acting as a structural element in this case.

Structural sealant glazing An external glazing system in which the glass is bonded to a carrier frame without mechanical retention. Often called structural silicone glazing when a silicone adhesive/sealant is used.

Sunlight The portion of solar energy detectable by the human eye; it accounts for about 44 percent of the total radiation wavelength spectrum.

Supercooled Frozen into shape.

Tank A glass furnace.

Tempered glass Glass that has been thermally treated to some extent. The term includes heat strengthened *and* fully tempered glass.

Tensile strength The maximum amount of tensile stress that a material can be subjected to before failure. The definition of failure can vary according to material type, limit state and design methodology.

Thermal break A material with a low thermal conductance used to separate exterior and interior materials. The thermal break is intended to stop the transfer of heat.

Thermal stress The internal stresses created when glass is subjected to variations in temperature across its area. If the temperature differentials in the glass are excessive, the glass may crack. This is referred to as thermal breakage or fracture.

Thermal transmittance See U-value.

Thermoplastic Capable of being repeatedly softened by heat and hardened by cooling.

Time of loading The time period during which a load is applied.

Tin side The lower side of glass in the float process, i. e. the side that is in contact with the pool of molten tin.

Tinted glass Transparent float glass with a consistent colour throughout its depth.

Total heat gain The sum of the energy transmitted into the building.

Total heat loss The sum of the energy transmitted to the outdoors.

Total transmittance See solar factor.

Toughened glass Term used in the UK for fully tempered glass (see *Table 1.6*).

Transient analysis An analysis that accounts for the time-dependence of input parameters.

Translucent Transmitting light but obscuring clear vision.

Transmittance The fraction of radiant energy that passes through a given material.

Transparent Clear, permitting vision.

Transverse seam A seam joining two materials across the width of the finished product.

Two-part compound A product which is necessarily packaged in two separate containers. It is comprised of a base and the curing agent or accelerator. The two compounds are uniformly mixed just prior to its use.

U-value The amount of conductive heat energy transferred through 1 m^2 of a specific insulating glass unit for 1 K temperature difference between the indoor and outdoor air. It is the inverse of the R-value ($U = 1/R$). Synonym: thermal transmittance.

Ultimate elongation The elongation at the moment of rupture.

Uniaxial stress field The minor principal stress is equal to zero. An uniaxial stress field is encountered for instance in four-point bending tests.

Uniform lateral load Uniformly distributed out-of-plane load.

Uniform stress field A stress field where the stress is equal at all points on the surface (cf. non-uniform stress field).

Vinyl glazing Holding glass in place with extruded vinyl channel or roll-in type.

Viscosity A measure of the resistance of a fluid to deformation under shear stress. Viscosity describes a fluid's internal resistance to flow and may be thought of as a measure of fluid friction.

Visible light transmittance The percentage of light in the visible spectrum range of 390 to 780 nm that is directly transmitted through glass.

Weep hole Opening at the base of a cavity wall to collect moisture and dispense it or a breather tube put in sealant to relieve moisture.

Wire glass Glass having a layer of meshed wire completely embedded in the glass pane. It may have polished or patterned surfaces.

Main sources: [7, 177, 189, 293, 347]

Appendix

C Statistical Fundamentals

C.1 Statistical distribution functions

Table C.1: Continuous statistical distribution functions.

Type	PDF $f(x)$ CDF $F(x)$	Mean μ Variance σ^2
Normal	$f(x) = \frac{1}{\sigma\sqrt{2\pi}} \exp\left(-\frac{1}{2}\left(\frac{x-\mu}{\sigma}\right)^2\right)$ $F(x) = \int_{-\infty}^{x} f(x)\,\mathrm{d}x$	$\mu = \mu$ $\sigma^2 = \sigma^2$
Log-normal	$f(x) = \frac{1}{\zeta x\sqrt{2\pi}} \exp\left(-\frac{1}{2}\left(\frac{\ln x-\lambda}{\zeta}\right)^2\right)$ $F(x) = \int_{0}^{x} f(x)\,\mathrm{d}x$	$\mu = \exp\left(\lambda + \frac{\zeta^2}{2}\right)$ $\sigma^2 = \mu^2\left(\exp(\zeta^2) - 1\right)$
Uniform	$f(x) = \frac{1}{b-a}$ $F(x) = \frac{x-a}{b-a}$	$\mu = \frac{a+b}{2}$ $\sigma^2 = \frac{(b-a)^2}{12}$
Pareto	$f(x) = \frac{ab^a}{x^{a+1}}$ $F(x) = 1 - \left(\frac{b}{x}\right)^a$	$\mu = \frac{ab}{a-1}$ $\sigma^2 = \frac{ab^2}{(a-1)^2(a-2)}$
Weibull	$f(x) = \frac{\beta}{\theta}\left(\frac{x}{\theta}\right)^{\beta-1} \cdot \exp\left(-\left(\frac{x}{\theta}\right)^{\beta}\right)$ $F(x) = 1 - \exp\left(-\left(\frac{x}{\theta}\right)^{\beta}\right)$	$\mu = \theta \cdot \Gamma\left(1 + \frac{1}{\beta}\right)$ $\sigma^2 = \theta^2\left[\Gamma\left(1+\frac{2}{\beta}\right) - \Gamma^2\left(1+\frac{1}{\beta}\right)\right]$

Table C.2: Discrete statistical distribution functions.

Type	PDF $f(x)$ CDF $F(x)$	Mean μ Variance σ^2	Notes
Poisson	$f(x) = \frac{e^{-\lambda}\lambda^x}{x!}$	$\mu = \lambda$	$x = 0, 1, 2, \ldots$
	$F(x) = \sum_{i=0}^{x} \frac{e^{-\lambda}\lambda^i}{i!}$	$\sigma^2 = \lambda$	

C.2 The empirical probability of failure

For some parameter estimation methods, for instance the least squares method (see Section C.3), an empirical probability distribution function for test data is required. In general, the probability density function (PDF) of the discrete random variable X is defined as

$$\hat{f}(x) = \begin{cases} p_i & \text{for } x = x_i \quad (i = 1, 2, 3, \ldots) \\ 0 & \text{for all other } x \end{cases} \tag{C.1}$$

with p_i being the probability that the random variable X takes on the value x_i, which means

$$p_i = \frac{N_i}{N} \tag{C.2}$$

in which N_i is the number of occurrences of the value i (generally 1 for test results) and N the total number of observations. The corresponding empirical cumulative distribution function is:

$$\hat{F}(x) = P(X \leq x) = \sum_{x_i \leq x} f(x_i) \tag{C.3}$$

If test results are ordered such that i is the rank of the value x_i within all test results, the most obvious estimator is:

$$\hat{F}(x_i) = \frac{i}{N} \tag{C.4}$$

While this estimator is very straightforward, it has at least two disadvantages. Firstly, the highest value cannot be represented on probability graphs and causes numerical problems. Secondly, it is very unlikely that the value with $\hat{F} = 1.0$ will be observed within relatively small samples. The largest value observed will thus lie below 1.0.

Values on the ordinate of a probability graph are actually random variables with a distribution of their own, which has a strong formal similarity to a beta distribution. The expectation value (mean rank) of this beta-distributed variable for the i-th value is

$$\hat{F}(x_i) = \frac{i}{N+1} \tag{C.5}$$

and is independent of the observed values' distribution. The use of Equation (C.5) is recommended by many standard works on statistics. For large samples, the difference between $N+1$ and N becomes very small. If the median (median rank) of the beta distribution is used instead of the expectation value, the estimator becomes[1]:

$$\hat{F}(x_i) = \frac{i - 0.3}{N + 0.4} \tag{C.6}$$

There is no straightforward way of telling which estimator is more suitable. The difference for practical application is small. In order to ensure consistency with the European standard on the determination of the strength of glass EN 12603:2002 [106], Equation (C.6) was used within the present work.

C.3 Parameter estimation and fitting algorithms

C.3.1 Maximum likelihood method

The principle of the maximum likelihood method is that the parameters of the distribution function are fitted such that the probability (likelihood) of the observed random sample is maximized.

Let X be a random variable with the probability density function $f(x, \vec{\theta})$ where $\vec{\theta} = (\theta_1, \theta_2, \ldots, \theta_K)^{\mathrm{T}}$ are the unknown constant parameters that need to be estimated. With the vector $\vec{x} = (\hat{x}_1, \hat{x}_2, \ldots, \hat{x}_N)$ containing the random samples from which the distribution parameters $\vec{\theta}$ are to be estimated, the likelihood function $L(\vec{x} \mid \vec{\theta})$ is given by the following product:

$$L(\vec{x} \mid \vec{\theta}) = \prod_{i=1}^{N} f(\hat{x}_i, \vec{\theta}) \tag{C.7}$$

The logarithmic likelihood function Λ, which is much easier to work with than L, is:

$$\Lambda(\vec{x} \mid \vec{\theta}) = \ln L(\vec{x} \mid \vec{\theta}) = \sum_{i=1}^{N} \ln f(\hat{x}_i, \vec{\theta}) \tag{C.8}$$

The maximum likelihood estimators of the parameters $\theta_1, \theta_2, \ldots, \theta_K$ are obtained by solving the following optimization problem:

$$\min_{\vec{\theta}}(-L(\vec{x} \mid \vec{\theta})) \qquad \text{or} \qquad \min_{\vec{\theta}}(-\Lambda(\vec{x} \mid \vec{\theta})) \tag{C.9}$$

As can be seen from the equations, the maximum likelihood method is independent of any kind of ranks or plotting methods (cf. Section C.2). The maximum likelihood estimators have a higher probability of being close to the quantities to be estimated than the point estimators obtained with the method of moments. [168, 349]

C.3.2 Least squares method

To obtain the coefficient estimates, the least squares method minimizes the summed square of residuals. The residual for the i-th data point Δ_i is defined as the difference between the observed response value y_i and the fitted response value $\hat{y}_i$, and is identified

[1]This is a good approximation, the exact solution can only be found through the roots of a polynomial.

as the error associated with the data. The summed square of the residuals (error estimate) is given by:

$$S = \sum_{i=1}^{I} \Delta_i^2 = \sum_{i=1}^{I} (y_i - \hat{y}_i)^2 \tag{C.10}$$

in which I is the number of data points included in the fit. [250]

C.3.3 Method of moments

EN 12603:2002 [106], the European standard for the analysis of glass strength data, is based on the method of moments. For uncensored samples, the following Weibull parameter point estimates are given:

$$\hat{\theta} = \exp\left[\frac{1}{N}\sum_{i=1}^{N} \ln x_i + 0.5772\frac{1}{\hat{\beta}}\right] \quad ; \quad \hat{\beta} = N\kappa_N \cdot \left(\frac{s}{N-s}\sum_{i=s+1}^{N} \ln x_i - \sum_{i=1}^{s} \ln x_i\right)^{-1} \tag{C.11}$$

$\hat{\theta}$ and $\hat{\beta}$ are the point estimates for the shape and scale parameters respectively. N is the sample size, x_i is the i-th sample, and s is the largest integer $< 0.84N$. The factor κ_N is a function of N and is provided in a table (examples: $N = 5 \Rightarrow \kappa_N = 1.2674$, $N = 10 \Rightarrow \kappa_N = 1.3644$, $N = 20 \Rightarrow \kappa_N = 1.4192$).

While the maximum likelihood method and the least squares method can be used to estimate parameters of any model, the point estimators in Equation (C.11) can *only* be used to estimate the parameters of a two-parameter Weibull distribution. Their main advantage is their simplicity.

References

[1] AAMA CW-13-85. *AAMA CW-13-85: Structural Sealant Glazing Systems*. American Architectural Manufacturers Association (AAMA), Schaumburg, USA, 1985.

[2] Abrams, M. B., Green, D. J. and Glass, S. J. Fracture behavior of engineered stress profile soda lime silicate glass. *Journal of Non-Crystalline Solids*, 321(1-2):10–19, 2003.

[3] Adams, R. and Harris, J. The influence of local geometry on the strength of adhesive joints. *International Journal of Adhesion & Adhesives*, 7(2):69–80, 1987.

[4] Adams, R. and Wale, W. *Structural adhesive joints in Engineering*. Elsevier, London, UK, 1984.

[5] Adams, R. D., Comyn, J. and Wake, W. C. *Structural Adhesive Joints in Engineering*. 2nd Edition. Chapman and Hall, London, 1997.

[6] Addington, M. and Schodek, D. *Smart Materials and Technologies in Architecture*. Architectural Press, 2004. ISBN 0750662255.

[7] Amstock, J. S. *Handbook of Glass in Construction*. McGraw-Hill, 1997. ISBN 0-070-01619-4.

[8] AS 1288-2006. *Glass in buildings – Selection and installation*. Australia, 2006.

[9] AS website. Angstrom Sciences website. 2006. URL http://www.angstromsciences.com/.

[10] ASCE GFI 1989. *Guidelines for Failure Investigation*. Technical Council on Forensic Engineering. American Society of Civil Engineers, 1989. ISBN 0872627365.

[11] ASTM C 1036-2001. *Standard Specification for Flat Glass*. American Society for Testing Materials, 2001.

[12] ASTM C 1048-04. *Standard Specification for Heat-Treated Flat Glass – Kind HS, Kind FT Coated and Uncoated Glass*. American Society for Testing Materials, 2004.

[13] ASTM C 1172-03. *Standard Specification for Laminated Architectural Flat Glass*. American Society for Testing Materials, 2003.

[14] ASTM C 1376-03. *Standard Specification for Pyrolytic and Vacuum Deposition Coatings on Flat Glass*. American Society for Testing Materials, 2003.

[15] ASTM C 1401-02. *Standard Guide for Structural Sealant Glazing*. American Society for Testing Materials, 2002.

[16] ASTM C 1422-99(2005). *Standard Specification for Chemically Strengthened Flat Glass*. American Society for Testing Materials, 2005.

[17] ASTM C 1464-06. *Standard Specification for Bent Glass*. American Society for Testing Materials, 2006.

[18] ASTM C 1503-01. *Standard Specification for Silvered Flat Glass Mirror*. American Society for Testing Materials, 2001.

[19] ASTM D 1002-05. *Standard test method for apparent shear strength of single-lap-joint adhesively bonded metal specimen by tension loading (metal-to-metal)*. American Society for Testing Materials, 2005.

[20] ASTM D 3528-96. *Standard test method for strength properties of double lap shear adhesive joints by tension loading*. American Society for Testing Materials, 1996.

[21] ASTM E 1300-03. *Standard Practice for Determining Load Resistance of Glass in Buildings*.

American Society for Testing Materials, 2003.

[22] ASTM E 1300-04. *Standard Practice for Determining Load Resistance of Glass in Buildings*. American Society for Testing Materials, 2004.

[23] ASTM E 1300-94. *Standard practice for determining the minimum thickness and type of glass required to resist a specific load*. American Society for Testing Materials, 1994.

[24] Atkins, A. G. and Mai, Y.-W. *Elastic and Plastic Fracture*. Ellis Horwood, 1988. ISBN 0-132-48196-0.

[25] Auradou, H., Vandembroucq, D., Guillot, C. and Bouchaud, E. A probabilistic model for the stress corrosion fracture of glass. *Transactions, SMiRT 16, Washington DC*, August 2001.

[26] Ayyub, B. M. and McCuen, R. H. *Probability. Statistics and Reliability for Engineers and Scientists*. Chapman & Hall/CRC Press, 2003

[27] Bakioglu, M., Erdogan, F. and Hasselman, D. P. H. Fracture mechanical analysis of self-fatigue in surface compression strengthened glass plates. *Journal of Materials Science*, 11:1826–1834, 1976.

[28] Bando, Y., Ito, S. and Tomozawa, M. Direct Observation of Crack Tip Geometry of SiO2 Glass by High-Resolution Electron Microscopy. *Journal of the American Ceramic Society*, 67(3):C36–C37, 1984.

[29] Batdorf, S. B. and Crose, J. G. Statistical Theory for the Fracture of Brittle Structures Subjected to Nonuniform Polyaxial Stresses. *Journal of Applied Mechanics, Transactions ASME*, 41 Ser E(2):459–464, 1974.

[30] Batdorf, S. B. and Heinisch, H. L. Weakest Link Theory Reformulated for Arbitrary Fracture Criterion. *Journal of the American Ceramic Society*, 61(7-8):355–358, 1978.

[31] Beason, W. L. *A failure prediction model for window glass*. NTIS Accession no. PB81-148421, Texas Tech University, Institute for Disaster Research, 1980.

[32] Beason, W. L., Kohutek, T. L. and Bracci, J. M. Basis for ASTM E 1300 annealed glass thickness selection charts. *Journal of Structural Engineering*, 142(2):215–221, February 1998.

[33] Beason, W. L. and Morgan, J. R. Glass Failure Prediction Model. *Journal of Structural Engineering*, 110(2):197–212, 1984.

[34] Behr, R. Architectural glass for earthquake-resistant buildings. In *Proceedings of Glass Processing Days 2001, 18-21 June, Tampere, Finland*. 2001.

[35] Belis, J. *Kipsterkte van monolithische en gelamineerde glazen liggers*. Ph.D. thesis, Ghent University, Belgium, 2005. URL http://hdl.handle.net/1854/5844.

[36] Belis, J. and Van Impe, R. Buckling-related problems of glass beams. In *Glass & Interactive Building Envelopes – Final Report EU COST Action C13 (to be published)*. Delft University of Technology & Ecole polytechnique fédérale de Lausanne EPFL, 2006.

[37] Belis, J., Van Impe, R., De Meester, B., Lagae, G. and Katnam, K. B. Stability Approach of the Dimensioning of Glass Beams. In *Proceedings of the International Symposium on the Application of Architectural Glass ISAAG 2004, 15-16 November, Munich, Germany*. 2004.

[38] Belis, J., Van Impe, R., Lagae, G. and Vanlaere, W. Enhancement of the buckling strength of glass bemas by means of lateral restraints. *Structutral Engineering and Mechanics*, 15(5):495–511, 2003.

[39] Bennison, S. J., Jagota, A. and Smith, C. A. Fracture of Glass/Poly(vinyl butyral) (Butacite®) Laminates in Biaxial Flexure. *Journal of the American Ceramic Society*, 82(7):1761–1770, 1999.

[40] Bennison, S. J., Smith, C. A., Van Duser, A. and Jagota, A. Structural Performance of Laminated Glass Made with a "Stiff" Interlayer. In *Glass in Buildings, ASTM STP 1434, V. Block (editor)*. 2002.

[41] Bernard, F. *Sur le dimensionnement des structures en verre trempé: étude des zones de connexion*. Ph.D. thesis, LMT Cachan, France, 2001.

[42] Bernard, F., Gy, R. and Daudeville, L. Finite Element Computation of Residual Stresses Near Holes in Tempered Glass Plates. *Glass Technology*, 43C, 2002.

[43] Blandini, L. *Structural use of adhesives in Glass Shells*. Ph.D. thesis, D 93 University of Stuttgart, Verlag Grauer, Stuttgart, 2005.

[44] Blank, K. *Dickenbemessung von vierseitig gelagerten rechteckigen Glasscheiben unter gleichförmiger Flächenlast*. Technical Report, Institut für Konstruktiven Glasbau, Gelsenkirchen, 1993.

[45] Bradt, R. C. *Fractography of Glass*. Kluwer Academic / Plenum Publishers, 1995. ISBN 0306448807.

[46] BRL-A 2005. *Bauregelliste A*. German building regulation, Deutsches Institut für Bautechnik (DIBt), 2005.

[47] Brown, W. G. A Load Duration Theory for Glass Design. *Publication NRCC 12354*, 1972.

[48] Brown, W. G. *A practicable Formulation for the Strength of Glass and its Special Application to Large Plates*. Publication No. NRC 14372, National Research Council of Canada, Ottawa, 1974.

[49] BS 6180:1999. *Barriers in and about buildings –*

Code of practice. British Standard Institute BSI, December 1999.

[50] BS 6206:1981. *Specification for impact performance requirements for flat safety glass and safety plastics for use in buildings*. British Standard Institute BSI, November 1981.

[51] BS 6262-3:2005. *Code of practice for glazing for buildings – Part 3: Code of practice for fire, security and wind loading*. British Standards Institution BSI, October 2005.

[52] BS 6262-4:1994. *Code of practice for glazing for buildings – Part 4. Safety related to human impact*. British Standard Institute BSI, August 1994.

[53] BS 6262-6:2005. *Code Of Practice For Glazing For Buildings – Part 6: Code Of Practice For Special Applications*. British Standards Institution BSI, October 2005. ISBN 0580467716.

[54] BS 6399-1:1996. *Loading for Buildings; Part 1. Code of practice for dead and imposed loads*. British Standard Institute BSI, September 1996.

[55] Bucak, O. and Hagl, A. Kleben im Bauwesen – gestern, heute, morgen499-507. *Stahlbau*, 75(6):499–507, 2006.

[56] Bucak, O. and Ludwig, J. J. Stand der Untersuchungen an grossformatigen TVG-Scheiben zur Erlangung einer bauaufsichtlichen Zulassung. In *Bauen mit Glas - VDI-Tagung in Baden-Baden 2000*, pp. 677–706. VDI-Verlag, Düsseldorf, 2000.

[57] Bunker, B. C. Molecular mechanisms for corrosion of silica and silicate glasses. *Journal of Non-Crystalline Solids*, 179:300–308, November 1994.

[58] Calderone, I. The Dangers of Using a Probabilistic Approach for Glass Design. In *Proceedings of Glass Processing Days 2005, 17-20 June, Tampere, Finland*, pp. 364–366. 2005.

[59] Calderone, I. J. The Fallacy of the Weibull Distribution for Window Glass Design. In *Proceedings of Glass Processing Days 2001, 18-21 June, Tampere, Finland*, pp. 293–297. 2001.

[60] CAN/CGSB 12.20-M89. *Structural Design of Glass for Buildings*. Canadian General Standards Board, December 1989.

[61] Carré, H. *Le verre trempé un nouveau matériau de structure*. Ph.D. thesis, Centre scientifique et technique du bâtiment - CSTB, Paris, 1997.

[62] Carré, H. and Daudeville, L. Numerical Simulation of Soda-Lime Silicate Glass Tempering. *Journal de Physique IV*, 6:175–185, 1996.

[63] Carré, H. and Daudeville, L. Thermal tempering simulation of glass plates: inner and edge residual stresses. *Journal of Thermal Stresses*, 21:667–689, 1998.

[64] Carré, H. and Daudeville, L. Load-Bearing Capacity of Tempered Structural Glass. *ASCE Journal of engineering mechanics*, 125, No. 8:914–921, 1999.

[65] Charles, R. J. and Hillig, W. B. The kinetics of glass failure by stress corrosion. In *Symposium on Mechanical Strength of Glass and Ways of Improving it, Charleroi, Belgium*. 1962.

[66] Charles, Y., Hild, F. and Roux, S. Long-term reliability of brittle materials: The issue of crack arrest. *Journal of Engineering Materials and Technology, Transactions of the ASME*, 125(3):333–340, 2003.

[67] Clark, A. B. and Irwin, G. R. Crack Propagation Behaviours. *Experimental Mechanics*, 6(6):321–330, 1966.

[68] Clarke, J. *Structural design of polymer composites – EUROCOMP design code and handbook*. Spon, London, 1996.

[69] Colvin, J. Look – No Frames! In *Proceedings of Glass Processing Days 2003, 15-18 June, Tampere, Finland*, pp. 208–210. 2003.

[70] Colvin, J. Thermal Stress Breakage in Building Glass, 2005. Workshop, Glass Processing Days 2005, 17-20 June, Tampere, Finland.

[71] Compagno, A. *Intelligente Glasfassaden – Material, Anwendung, Gestaltung / Intelligent Glass Façades – Material, Practice, Design*. Birkhäuser Verlag, 2002. ISBN 3-7643-6694-X.

[72] Conway, J. C. and Mecholsky, J. J. Use of Crack Branching Data for Measuring Near-Surface Residual Stresses in Tempered Glass. *Journal of the American Ceramic Society*, 72(9):1584–1587, 1989.

[73] Cook, R. D. *Finite Element Modelling For Stress Analysis*. Wiley, New York, 1995. ISBN 0471107743.

[74] CPSC 16-1201:2003. *CPSC 16 CFR, Part 1201: Safety standard for architectural glazing materials*. USA Consumer Product Safety Commission, Code of Federal Regulations, 2003.

[75] CSTB 3488. *Vitrages extérieurs collés – Cahier des prescriptions techniques*. Cahier du CSTB, Livraison 444, Cahier 3488, November 2003.

[76] CWCT 2003. *Use of glass overhead*. Technical Update No. 10, Centre for Window and Cladding Technology (CWCT), Bath, UK, 2003.

[77] CWCT 2004. *Safety & fragility of glazed roofing*. Technical Note No. 42, Centre for Window and Cladding Technology (CWCT), Bath, UK, 2000.

[78] De Gaetano, S. Design for Glazing Protection Against Terrorist Attacks. In *Proceedings of Glass Processing Days 2005, 17-20 June, Tampere, Finland*. 2005.

[79] DIBt 2001. *Merkblatt "Anforderungen an begehbare Verglasungen; Empfehlungen für das Zus-*

timmungsverfahren". Fassung März 2000, Mitteilungen des Deutschen Instituts für Bautechnik (DIBt), Berlin, 2001.

[80] DIN 1249-10:1990. *Flachglas im Bauwesen – Teil 10: Chemische und physikalische Eigenschaften*. DIN, 1990.

[81] DIN 18008-1:2006. *Glass in Building – Design and construction rules – Part 1: Terms and general bases*. German Draft Standard, 2006.

[82] DIN 18008-2:2006. *Glass in Building – Design and construction rules – Part 2: Linearly supported glazings*. German Draft Standard, 2006.

[83] DIN 18516-4:1990. *Außenwandbekleidungen, hinterlüftet – Einscheiben-Sicherheitsglas – Teil 4: Anforderungen, Bemessung, Prüfung*. DIN, 1990.

[84] DIN 4426:2001. *Sicherheitstechnische Anforderungen an Arbeitsplätze und Verkehrswege – Planung und Ausführung*. DIN, September 2001.

[85] DIN 52292-2:1986. *Teil 2: Prüfung von Glas und Glaskeramik. Bestimmung der Biegefestigkeit, Doppelring-Biegeversuch an plattenförmigen Proben mit grossen Prüfflächen*. DIN, 1986.

[86] DIN 53286:2006. *DIN 53287: Testing of adhesives for metals and adhesively bonded metal joints – Determination of the resistance to liquids*. 2006.

[87] DIN 54456:2006. *DIN 54456: Testing of structural adhesives – Test of resistance to climatic conditions*. 2006.

[88] DIN 55303-7:1996. *Statistische Auswertung von Daten – Teil 7: Schätz- und Testverfahren bei zweiparametriger Weibull-Verteilung*. DIN, 1996.

[89] DowCorning 2005. *Dow Corning Americas Technical Manual*. Dow Corning, 2005. URL www.dowcorning.com.

[90] DowCorning 2006. *Dow Corning: Silicone Structural Glazing Manual*, 2006. URL http://www.dowcorning.com/.

[91] DTU 39 - A1. *DTU 39 – Travaux de miroiterie – Vitrerie Partie 1 : Cahier des charges techniques – Amendement A1 sur la "casse thermique" des vitrages (NF P 78-201-1/A1)*. Association Française de Normalisation (AFNOR), May 1998.

[92] Duckworth, W., Shetty, D., Rosenfield, A. and Siskos, W. Influence of Stress Gradients on the Relationship Between Fracture Stress and Mirror Size for Float Glass. *Glass Technology*, 24(5):263–273, 1983.

[93] DuPont 2003. *SentryGlas Plus – The Interlayer for Structural Laminated Glass*. DuPont, 2003.

[94] Dwivedi, P. and Green, D. J. Determination of Subcritical Crack Growth Parameters by In Situ Observations of Identation Cracks. *Journal of the American Ceramic Society*, 78(8):2122–2128, 1995.

[95] EN 1051-1:2003. *Glass in building – Glass blocks and glass paver units – Part 1: Definitions and description*. CEN, 2003.

[96] EN 1051-2:2003. *Glass in building – Glass blocks and glass paver units – Part 2: Evaluation of conformity*. CEN, 2003.

[97] EN 1096-1:1998. *Glass in building – Coated glass – Part 1: Definitions and classification*. CEN, January 1999.

[98] EN 1096-2:2001. *Glass in building – Coated glass – Part 2: Requirements and test methods for class A, B and S coatings*. CEN, 2001.

[99] EN 1096-3:2001. *Glass in building – Coated glass – Part 3: Requirements and test methods for class C and D coatings*. CEN, 2001.

[100] EN 1096-4:2004. *Glass in building – Coated glass – Part 4: Evaluation of conformity / Product standard*. CEN, 2004.

[101] EN 12150-1:2000. *Glass in building – Thermally toughened soda lime silicate safety glass – Part 1: Definition and description*. CEN, 2000.

[102] EN 12150-2:2004. *Glass in building – Thermally toughened soda lime silicate safety glass – Part 2: Evaluation of conformity / Product standard*. CEN, 2004.

[103] EN 12337-1:2000. *Glass in building – Chemically strengthened soda lime silicate glass – Part 1: Definition and description*. CEN, 2000.

[104] EN 12337-2:2004. *Glass in building – Chemically strengthened soda lime silicate glass – Part 2: Evaluation of conformity / Product standard*. CEN, 2004.

[105] EN 12600:2002. *Glass in building – Pendulum test – Impact test method and classification for flat glass*. CEN, 2002.

[106] EN 12603:2002. *Glass in building – Procedures for goodness of fit and confidence intervals for Weibull distributed glass stength data*. CEN, 2002.

[107] EN 1279-1:2004. *Glass in building – Insulating glass units – Part 1: Generalities, dimensional tolerances and rules for the system description*. CEN, 2004.

[108] EN 1279-2:2002. *Glass in building – Insulating glass units – Part 2: Long term test method and requirements for moisture penetration*. CEN, 2002.

[109] EN 1279-3:2002. *Glass in building – Insulating glass units – Part 3: Long term test method and requirements for gas leakage rate and for gas concentration tolerances*. CEN, 2002.

[110] EN 1279-4:2002. *Glass in building – Insulating glass units – Part 4: Methods of test for the physical attributes of edge seals.* CEN, 2002.

[111] EN 1279-5:2005. *Glass in building – Insulating glass units – Part 5: Evaluation of conformity.* CEN, 2005.

[112] EN 1279-6:2002. *Glass in building – Insulating glass units – Part 6: Factory production control and periodic tests.* CEN, 2002.

[113] EN 1288-1:2000. *Glass in building – Determination of the bending strength of glass – Part 1: Fundamentals of testing glass.* CEN, 2000.

[114] EN 1288-2:2000. *Glass in building – Determination of the bending strength of glass – Part 2: Coaxial double ring test on flat specimens with large test surface areas.* CEN, 2000.

[115] EN 1288-3:2000. *Glass in building – Determination of the bending strength of glass – Part 3: Test with specimen supported at two points (four-point bending).* CEN, 2000.

[116] EN 1288-5:2000. *Glass in building – Determination of the bending strength of glass – Part 5: Coaxial double ring test on flat specimens with small test surface areas.* CEN, 2000.

[117] EN 13024-1:2002. *Glass in building – Thermally toughened borosilicate safety glass – Part 1: Definition and description.* CEN, 2002.

[118] EN 13024-2:2004. *Glass in building – Thermally toughened borosilicate safety glass – Part 2: Evaluation of conformity / Product standard.* CEN, 2004.

[119] EN 13123-2:2004. *Windows, doors, and shutters – Explosion resistance – Requirements and classification – Part 2: Range test.* CEN, 2004.

[120] EN 13124-1:2001. *Windows, doors and shutters – Explosion resistance; Test method – Part 1: Shock tube.* CEN, 2001.

[121] EN 13124-2:2004. *Windows, doors and shutters – Explosion resistance – Test method – Part 2: Range test.* CEN, 2004.

[122] EN 13541:2005. *Glass in building – Security glazing – Testing and classification of resistance against explosion pressure.* CEN, 2005.

[123] EN 14179-1:2005. *Glass in building – Heat soaked thermally toughened soda lime silicate safety glass – Part 1: Definition and description.* CEN, 2005.

[124] EN 14179-2:2005. *Glass in building – Heat soaked thermally toughened soda lime silicate safety glass – Part 2: Evaluation of conformity / Product standard.* CEN, 2005.

[125] EN 14321-1:2005. *Glass in building – Thermally toughened alkaline earth silicate safety glass – Part 1: Definition and description.* CEN, 2005.

[126] EN 14321-2:2005. *Glass in building – Thermally toughened alkaline earth silicate safety glass – Part 2: Evaluation of conformity / Product standard.* CEN, 2005.

[127] EN 14449:2005. *Glass in building – Laminated glass and laminated safety glass – Evaluation of conformity / Product standard.* CEN, 2005.

[128] EN 1465:1995. *EN 1465: Adhesives – Determination of tensile lap-shear strength of rigid-to-rigid bonded assemblies.* CEN, 1995.

[129] EN 1748-1-1:2004. *Glass in building – Special basic products – Borosilicate glasses – Part 1-1: Definitions and general physical and mechanical properties.* CEN, 2004.

[130] EN 1748-1-2:2004. *Glass in building – Special basic products – Borosilicate glasses – Part 1-2: Evaluation of conformity / Product standard.* CEN, 2004.

[131] EN 1748-2-1:2004. *Glass in building – Special basic products – Glass ceramics – Part 2-1 Definitions and general physical and mechanical properties.* CEN, 2004.

[132] EN 1748-2-2:2004. *Glass in building – Special basic products – Part 2-2: Glass ceramics – Evaluation of conformity / Product standard.* CEN, 2004.

[133] EN 1863-1:2000. *Glass in building – Heat strengthened soda lime silicate glass – Part 1: Definition and description.* CEN, 2000.

[134] EN 1863-2:2004. *Glass in building – Heat strengthened soda lime silicate glass – Part 2: Evaluation of conformity / Product standard.* CEN, 2004.

[135] EN 1990:2002. *Grundlagen der Tragwerksplanung.* CEN, 2002.

[136] EN 1991-1-1:2002. *Eurocode 1: Einwirkungen auf Tragwerke; Teil 1-1: Allgemeine Einwirkungen auf Tragwerke.* CEN, 2002.

[137] EN 1991-1-7:2006. *Eurocode 1. Actions on structures. General actions – Accidental actions.* CEN, 2006.

[138] EN 1991-2-3:1996. *Eurocode 1: Einwirkungen auf Tragwerke; Teil 2-3: Schneelasten.* CEN, 1996.

[139] EN 1991-2-4:1995. *Eurocode 1: Einwirkungen auf Tragwerke; Teil 2-4: Windlasten.* CEN, December 1996.

[140] EN 1991-2-5:1997. *Eurocode 1: Einwirkungen auf Tragwerke; Teil 2-5: Temperatureinwirkungen.* CEN, January 1999.

[141] EN 1991-2-7:1998. *Eurocode 1: Einwirkungen auf Tragwerke; Teil 2-7: Aussergewöhnliche Einwirkungen.* CEN, July 2000.

[142] EN 1993-1-1:1993. *Eurocode 3: Bemessung und Konstruktion von Stahlbauten; Allgemeine*

Bemessungsregeln, Bemessungsregeln für den Hochbau. CEN, January 1993.

[143] EN 1993-1-1:2005. *Eurocode 3: Design of steel structures – Part 1-1: General rules and rules for buildings*. CEN, 2005.

[144] EN 2243:2006. *EN 2243: Aerospace series – Non-metallic materials – Structural adhesives – Test method*. CEN, 2006.

[145] EN 2667:1997. *EN 2667: Aerospace series – Non-metallic materials; foaming structural adhesives, test methods*. CEN, 1997.

[146] EN 356:1999. *Glass in building – Security glazing – Testing and classification of resistance against manual attack*. CEN, 2000.

[147] EN 410:1998. *Glass in building – Determination of luminous and solar characteristics of glazing*. CEN, 1998.

[148] EN 572-1:2004. *Glass in building – Basic soda lime silicate glass products – Part 1: Definitions and general physical and mechanical properties*. CEN, 2004.

[149] EN 572-2:2004. *Glass in building – Basic soda lime silicate glass products – Part 2: Float glass*. CEN, 2004.

[150] EN 572-3:2004. *Glass in building – Basic soda lime silicate glass products – Part 3: Polished wire glass*. CEN, 2004.

[151] EN 572-4:2004. *Glass in building – Basic soda lime silicate glass products – Part 4: Drawn sheet glass*. CEN, 2004.

[152] EN 572-5:2004. *Glass in building – Basic soda lime silicate glass products – Part 5: Patterned glass*. CEN, 2004.

[153] EN 572-6:2004. *Glass in building – Basic soda lime silicate glass products – Part 6: Wired patterned glass*. CEN, 2004.

[154] EN 572-7:2004. *Glass in building – Basic soda lime silicate glass products – Part 7: Wired or unwired channel shaped glass*. CEN, 2004.

[155] EN 572-8:2004. *Glass in building – Basic soda lime silicate glass products – Part 8: Supplied and final cut sizes*. CEN, 2004.

[156] EN 572-9:2004. *Glass in building – Basic soda lime silicate glass products – Part 9: Evaluation of conformity / Product standard*. CEN, 2004.

[157] EN 673:1997. *Glass in building – Determination of thermal transmittance (U value) – Calculation method*. CEN, 1997.

[158] EN ISO 291:2005. *EN ISO 291: Plastics – Standard atmospheres for conditioning and testing*. CEN, 2005.

[159] EN ISO 527:1996. *EN ISO 527: Plastics – Determination of tensile properties*. CEN, 1996.

[160] EN ISO 899:2003. *EN ISO 899: Plastics – Determination of creep behaviour*. CEN, 2003.

[161] EN ISO 9142:2004. *EN ISO 9142: Adhesives – Guide to the selection of standard laboratory ageing conditions for testing bonded joints*. CEN, 2004.

[162] Englhardt, O. and Bergmeister, K. Structural Behaviour of Plane Glass Surface Structures Under Compression StressNumerical and Experimental Investigation. In *Proceedings of Glass Processing Days 2005, 17-20 June, Tampere, Finland*. 2005.

[163] EOTA 1998. *Leitlinie für die europäische technische Zulassung für geklebte Glaskonstruktionen (Structural Sealant Glazing Systems – SSGS)*. ETAG Nr. 002 (Fassung Juli 1998), EOTA - European Organisation for Technical Approvals, Brüssel, 1998.

[164] Evans, A. G. A General Approach for the Statistical Analysis of Multiaxial Fracture. *Journal of the American Ceramic Society*, 61(7-8):302–308, 1978.

[165] Evans, A. G. and Wiederhorn, S. M. Proof testing of ceramic materials – an analytical basis for failure prediction. *International Journal of Fracture*, 10(3):379–392, 1974.

[166] Exner, G. Erlaubte Biegespannung in Glasbauteilen im Dauerlastfall: Ein Vorhersagekonzept aus dynamischen Labor-Festigkeitsmessungen. *Glastechnische Berichte*, 56(11):299–312, 1983.

[167] Exner, G. Abschätzung der erlaubten Biegespannung in vorgespannten Glasbauteilen. *Glastechnische Berichte*, 59(9):259–271, 1986.

[168] Faber, M. *Risk and Safety in Civil Engineering*. Lecture Notes, Swiss Federal Institute of Technology ETH, Zürich, 2003.

[169] Fink, A. *Ein Beitrag zum Einsatz von Floatglas als dauerhaft tragender Konstruktionswerkstoff im Bauwesen*. Ph.D. thesis, Technische Hochschule Darmstadt, 2000.

[170] Fischer-Cripps, A. C. and Collins, R. E. Architectural glazings: Design standards and failure models. *Building and Environment*, 30(1):29–40, 1995.

[171] Freund, L. B. *Dynamic Fracture Mechanics*. Cambridge University Press, 1990. ISBN 0-521-30330-3.

[172] Gehrke, E., Ullner, C. and Hähnert, M. Effect of corrosive media on crack growth of model glasses and commercial silicate glasses. *Glastechnische Berichte*, 63(9):255–265, 1990.

[173] Gehrke, E., Ullner, C. and Hähnert, M. Fatigue limit and crack arrest in alkali-containing silicate glasses. *Journal of Materials Science*, 26:5445–5455, 1991.

[174] Georg, A. A chemical solar eclipse in your window. Press Release, Fraunhofer-Institut für

Solare Energiesysteme ISE, Freiburg, Germany, 2002.

[175] GGF 2000. *Non-vertical overhead glazing: Guide to the selection of glass from the point of view of safety*. Architecural glazing systems, Section 7, Glass & Glazing Federation, London, 2000.

[176] Glass Guide 2000. *Glass Guide*. Saint Gobain Glass UK, 2000.

[177] GlassOnWeb. GlassOnWeb – Glass manual. 2006. URL http://www.glassonweb.com/glassmanual/.

[178] Gough, M. and Yandzio, E. *Protection of Buildings Against Explosions*. The Steel Construction Institute, 1999. ISBN 1859420893.

[179] Green, D. J., Dwivedi, P. J. and Sglavo, V. M. Characterization of subcritical crack growth in ceramics using indentation cracks. *British Ceramic Transactions*, 98(6):291–295, 1999.

[180] Grenet, L. Mechanical Strength of Glass. *Enc. Industr. Nat. Paris*, 5(4):838–848, 1899.

[181] Griffith, A. A. The Phenomena of Rupture and Flow in Solids. *Philosophical Transactions, Series A*, 221:163–198, 1920.

[182] GS-BAU-18:2001. *Grundsätze für die Prüfung und Zertifizierung der bedingten Betretbarkeit oder Durchsturzsicherheit von Bauteilen bei Bau- oder Instandhaltungsarbeiten (Prüfgrundsätze GS-BAU-18)*. Technical Report, Hauptverband der gewerblichen Berufgenossenschaften HVBG, February 2001.

[183] GSA 2003. *GSA Standard Test Method for Glazing and Window Systems Subject to Dynamic Overpressure Loadings*. U.S. General Services Administration, 2003.

[184] Güsgen, J. *Bemessung tragender Bauteile aus Glas*. Ph.D. thesis, RWTH Aachen / Shaker Verlag, 1998.

[185] Guin, J.-P. and Wiederhorn, S. Crack growth threshold in soda lime silicate glass: role of hold-time. *Journal of Non-Crystalline Solids*, 316(1):12–20, 2003.

[186] Gy, R. Stress corrosion of silicate glass: a review. *Journal of Non-Crystalline Solids*, 316(1):1–11, 2003.

[187] Habenicht, G. *Kleben. Grundlagen, Technologie, Anwendungen*. Springer, Berlin, Germany, 1997.

[188] Hagl, A. Die Innovation – Kleben – Aktuelles aus der Arbeitsgruppe Kleben des Fachverbandes Konstruktiver Glasbau – FKG. *Stahlbau*, 75(6):508–520, 2006.

[189] Haldimann, M. *Fracture Strength of Structural Glass Elements – Analytical and numerical modelling, testing and design*. Thèse EPFL No 3671, Ecole polytechnique fédérale de Lausanne (EPFL), 2006. URL http://icom.epfl.ch/publications/pubinfo.php?pubid=561.

[190] Haldimann, M. and Crisinel, M. Improving testing procedures to obtain more reliable strength data for structural glass design. In *Proceedings of Glass Processing Days 2007, 15-18 June, Tampere, Finland, 2007*. 2007.

[191] Hamm, J. *Tragverhalten von Holz und Holzwerkstoffen im statischen Verbund mit Glas*. Ph.D. thesis, Ecole polytechnique fédérale de Lausanne (EPFL), Switzerland, 1999.

[192] Hand, R. J. Stress intensity factors for surface flaws in toughened glasses. *Fatigue and Fracture of Engineering Materials and Structures*, 23(1):73–80, 2000.

[193] Harris, R., Lilly, R. and Willmott, T. A Consultants Toolbox for Investigating Nickel Sulfide Failures in Toughened Glass. In *Proceedings of Glass Processing Days 2003, 15-18 June, Tampere, Finland*, pp. 680–682. 2003.

[194] Hess, R. *Glasträger*. Forschungsbericht des Instituts für Hochbautechnik, vdf Hochschulverlag AG der ETH Zürich, Zürich, 2000.

[195] Hilti HY-50 2003. *Konstruktiver Glasbau mit Hilti Injektionsmörtel HIT-HY 50*. Hilti Deutschland GmbH, 2003.

[196] Hirt, M. A. and Bez, R. Stahlbau: Grundbegriffe und Bemessungsverfahren. 1998.

[197] Hénaux, S. and Creuzet, F. Kinetic fracture of glass at the nanometer scale. *Journal of Materials Science Letters*, 16(12):1008–1011, 1997.

[198] HomeOffice 2005. *Design for Glazing protection*. Guideline document with restricted availability, Home Office Scientific Development Branch, UK, 2005.

[199] Hull, D. *Fractography: Observing, Measuring and Interpreting Fracture Surface Topography*. Cambridge University Press, 1999. ISBN 0521646847.

[200] IBC 2003. *International Building Code (IBC)*. Thomson Delmar Learning, 2003. ISBN 1-892-39556-8.

[201] Inglis, C. E. Stresses in a plate due to the presence of cracks and sharp corners. *Transactions of the Institution of Naval Architects*, 55(219), 1913.

[202] Interpane Glas Industrie AG 1997. *Gestalten mit Glas*. 5. Auflage, Lauenförde, Interpane Glas Industrie AG, 1997.

[203] Irwin, G. Analysis of Stresses and Strains near the End of a Crack Traversing a Plate. *Journal of Applied Mechanics*, 24:361–364, 1957.

[204] Irwin, G. R., Krafft, J. M., Paris, P. C. and Wells, A. A. *Basic Aspects of Crack Growth and Fracture*. NRL Report 6598, Naval Research Labo-

ratory, Washington D.C., 1967.

[205] ISO 12543-1:1998. *Glass in building – Laminated glass and laminated safety glass – Part 1: Definitions and description of component parts*. ISO, 1998.

[206] ISO 12543-2:2004. *Glass in building – Laminated glass and laminated safety glass – Part 2: Laminated safety glass*. ISO, 2004.

[207] ISO 12543-3:1998. *Glass in building – Laminated glass and laminated safety glass – Part 3: Laminated glass*. ISO, 1998.

[208] ISO 12543-4:1998. *Glass in building – Laminated glass and laminated safety glass – Part 4: Test methods for durability*. ISO, 1998.

[209] ISO 12543-5:1998. *Glass in building – Laminated glass and laminated safety glass – Part 5: Dimensions and edge finishing*. ISO, 1998.

[210] ISO 12543-6:1998. *Glass in building – Laminated glass and laminated safety glass – Part 6: Appearance*. ISO, 1998.

[211] ISO 31-0:1992. *Quantities and units – Part 0: General principles*. ISO, 1992.

[212] ISO 31-0:1992/Amd.2:2005. *Quantities and units – Part 0: General principles – Amendment 2*. ISO, 2005.

[213] ISO 31-11:1992. *Quantities and units – Part 11: Mathematical signs and symbols for use in the physical sciences and technology*. ISO, 1992.

[214] ISO 31-3:1992. *Quantities and units – Part 3: Mechanics*. ISO, 1992.

[215] ISO 9047:2003. *ISO 9047:2003, Building construction. Jointing products. Determination of adhesion/cohesion properties of sealants at variable temperatures*. 2003.

[216] ISO/DIS 16933:2004. *Glass in building – Explosion-resistant security glazing – Test and classification by arena air-blast loading*. CEN, 2004.

[217] ISO/DIS 16934:2004. *Glass in building – Explosion-resistant security glazing – Test and classification by shock tube loading*. CEN, 2004.

[218] IStructE 1995. *The structural engineer's response to blast loading*. The Institution of Structural Engineers, London, November 1995. ISBN 1-874266-20-4.

[219] IStructE 1999. *Structural use of glass in buildings*. The Institution of Structural Engineers, London, December 1999. ISBN 1-874266-51-4.

[220] IStructE Adh 1999. *Guide to the Structural Use of Adhesives*. The Institution of Structural Engineers, London, 1999.

[221] Johar, S. *Dynamic fatigue of flat glass – Phase II*. Technical Report, Ontario Research Foundation. Department of Metals, Glass and Ceramics, Missisagua, Canada, 1981.

[222] Johar, S. *Dynamic fatigue of flat glass – Phase III*. Technical Report, Ontario Research Foundation. Department of Metals, Glass and Ceramics, Missisagua, Canada, 1982.

[223] Johnson, N. International standards for blast resistant glazing. In *Proceedings of Glass in Buildings 2, 7-8 April 2005, Centre for Window & Cladding Technology (CWCT), Bath, UK*. 2005.

[224] Kasper, R. *Tragverhalten von Glasträgern*. Ph.D. thesis, RWTH Aachen / Shaker Verlag, 2005. URL http://www.shaker.de/.

[225] Keiller, A., Walker, A., Ledbetter, S. and Wolmuth, W. *Guidance on glazing at height*. CIRIA Publication C632, CIRIA, London, 2005.

[226] Kerkhof, F. Bruchmechanik von Glas und Keramik. *Sprechsaal*, 110:392–397, 1977.

[227] Kerkhof, F., Richter, H. and Stahn, D. Festigkeit von Glas – Zur Abhängigkeit von Belastungsdauer und -verlauf. *Glastechnische Berichte*, 54(8):265–277, 1981.

[228] Kinloch, A. *Durability of Structural Adhesives*. Applied Science Publishers, London, 1983.

[229] Knippers, J. and Peters, S. GFK-Glas-Verbundsysteme, in Transluzente Materialien. pp. 36–38, 2003.

[230] Ko, N., You, K. and Kim, Y. The effect of non-gaussian local wind pressures on a side face of a square building. *Journal of Wind Engineering and Industrial Aerodynamics*, 93(5):383–397, 2005.

[231] Kott, A. *Zum Trag- und Resttragverhalten von Verbundsicherheitsglas*. Ph.D. thesis, Eidgenössische Technische Hochschule Zürich (ETH) - Institut für Baustatik und Konstruktion (IBK), Zurich, Switzerland, 2006.

[232] Kott, A. and Vogel, T. Safety of laminated glass structures after initial failure. In *Proceedings of the IABSE Symposium Shanghai, September 22-24*. 2004.

[233] Kreher, K. *Tragverhalten und Bemessung von Holz-Glas-Verbundträgern unter Berücksichtigung der Eigenspannungen im Glas*. Thèse EPFL 2999, Ecole polytechnique fédérale de Lausanne (EPFL), 2004.

[234] Kutterer, M. Verbundglasplatten – Schubverbund und Membrantragwirkung – Teil 1. *Stahlbau*, 74(1):39–46, January 2005.

[235] Lampert, C. M. Chromogenic smart materials. *Materials Today*, 7(3):28–35, 2004.

[236] Laufs, W. *4. Forschungsbericht Fachverband Konstruktiver Glasbau, Arbeitskreis Punktgestützte Gläser*. Technical Report, Fachverband Konstruktiver Glasbau, Arbeitskreis Punktgestützte Gläser, Aachen, 2000.

[237] Laufs, W. *Ein Bemessungskonzept zur Festigkeit thermisch vorgespannter Gläser*. Ph.D. thesis, RWTH Aachen / Shaker Verlag, 2000.

[238] Lawn, B. *Fracture of brittle solids*. Second Edition. Cambridge University Press, 1993.

[239] Levengood, W. C. Effect of origin flaw characteristics on glass strength. *Journal of Applied Physics*, 29(5):820–826, 1958.

[240] Lindner, J. and Holberndt, T. Zum Nachweis von stabilitätsgefährdeten Glasträgern unter Biegebeanspruchung. *Stahlbau*, 75(6):488–498, 2006.

[241] Lotz, S. *Untersuchung zur Festigkeit und Langzeitbeständigkeit adhäsiver Verbindungen zwischen Fügepartnern aus Floatglas*. Ph.D. thesis, Universität Kaiserslautern, Fachbereich Maschinenwesen, 1995.

[242] Luible, A. *Lasteinleitungsversuche in Glaskanten*. Rapport ICOM 463, Ecole polytechnique fédérale de Lausanne (EPFL), 2004.

[243] Luible, A. *Stabilität von Tragelementen aus Glas*. Thèse EPFL 3014, Ecole polytechnique fédérale de Lausanne (EPFL), 2004. URL http://icom.epfl.ch/publications/pubinfo.php?pubid=499.

[244] Luible, A. and Crisinel, M. Buckling Strength of Glass Elements in Compression. *Structural Engineering International*, 14(2):120–125, 2004.

[245] Luible, A. and Crisinel, M. Plate Buckling of Glass Panels. In *Proceedings of Glass Processing Days 2005, 17-20 June, Tampere, Finland*, pp. 476–479. 2005.

[246] Luible, A. and Crisinel, M. Design of Glass Beams Subjected to Lateral Torsional Buckling. In *Proceedings of the IABSE Symposium 2006 "Responding to Tomorrow's Challenges in Structural Engineering", 13-15 Sept., Budapest, Hungary*. 2006.

[247] Luible, A. and Crisinel, M. Lateral Torsional Buckling of Glass Beams. In *Proceedings of the 11th International Conference on Metal Structures (ICMS'2006), 14-16 June, Rzeszow, Poland*. 2006.

[248] Maniatis, I. *Numerical and Experimental Investigations on the Stress Distribution of Bolted Glass Connections under In-Plane Loads*. Ph.D. thesis, Technische Universität München (TUM), 2006.

[249] Marlière, C., Prades, S., Célarié, F., Dalmas, D., Bonamy, D., Guillot, C. and Bouchaud, E. Crack fronts and damage in glass at the nanometre scale. *Journal of Physics: Condensed Matter*, 15:2377–2386, 2003.

[250] Matlab Doc 2005. *Documentation for Matlab 7.1*. The MathWorks, Inc., 2005.

[251] Mays, G. and Hutchinson, A. *Adhesives in civil engineering*. Cambridge University Press, Cambridge, 1992.

[252] Mays, G. C. and Smith, P. D. *Blast effects on Buildings*. Thomas Telford Publishers, 1995. ISBN 0727720309.

[253] McClelland, N. Not Just Glass Fins – MIT Brain and Cognitive Science Project. In *Proceedings of Glass Processing Days 2005, 17-20 June, Tampere, Finland*, pp. 480–484. 2005.

[254] Menčík, J. *Strength and Fracture of Glass and Ceramics, Glass Science and Technology*, volume 12. Elsevier, 1992.

[255] Michalske, T. A. and Freiman, S. W. A Molecular Mechanism for Stress Corrosion in Vitreous Silica. *Journal of the American Ceramic Society*, 66(4):284–288, 1983.

[256] Montgomery, D. C. and Runger, G. C. *Applied Statistics and Probability for Engineers*. John Wiley & Sons, New York, 1994.

[257] Mott, N. F. Brittle fracture in mild steel plates. *Engineering*, 165(16), 1948.

[258] Neugebauer, J. A Special Fixation With Which The Broken Laminated Safety Glass Is Prevent From Falling Down. In *Proceedings of Glass Processing Days 2005, 17-20 June, Tampere, Finland*, pp. 105–109. 2005.

[259] Nghiem, B. *Fracture du verre et heterogeneites a l'echelle sub-micronique*. Ph.D Thesis Nr 98 PA06 6260, Université Paris-VI, 1998.

[260] Niedermaier, P. Shear strength of glass panel elements in combination with timber frame constructions. In *Proceedings of Glass Processing Days 2003, 15-18 June, Tampere, Finland*, pp. 262–264. 2003.

[261] Noon, R. K. *Forensic Engineering Investigation*. CRC Press, 2001. ISBN 0849309115.

[262] Oakley, D. R. Crack branching in float glass subjected to biaxial loading. *Journal of Non-Crystalline Solids*, 196:139–143, 1996.

[263] Overend, M. *The Appraisal of Structural Glass Assemblies*. Ph.D. thesis, University of Surrey, March 2002.

[264] Overend, M. Optimising connections in structural glass. In *Proceedings of Glass in Buildings 2, 7-8 April 2005*. Centre for Window & Cladding Technology (CWCT), Bath, UK, 2005.

[265] Overend, M., De Gaetano, S. and Haldimann, M. Diagnostic Interpretation of Glass Failure. *Structural Engineering International SEI*, 17(2):151–158, 2007.

[266] Overend, M. and Koumpli, E. Smart Transparency – The use of chromogenic switchable glazing in responsive facades. In *International Conference on "Transparency and Architecture: Challenging the Limits", Thessaloniki, Greece*. 2006.

[267] Overend, M. and Maniatis, I. The influence of finite element modelling parameters on the stress distribution in bolted glass connections. In *11th International Conference on Civil, Structural and Environmental Engineering Computing, Malta*. 2007.

[268] Overend, M., Vassallo, C. and Camillieri, K. The Design, Assembly & Performance of Glass Columns. In *Proceedings of Glass Processing Days 2005, 17-20 June, Tampere, Finland*, pp. 287–291. 2005.

[269] Overend, M. and Zammit, K. Wind loading on cladding and glazed façades. In *Proceedings of the 2nd International Symposium on the Application of Architectural Glass ISAAG 2006, Munich, Germany*. 2006.

[270] Palumbo, D., Palumbo, M. and Mazzucchelli, M. A new roof for the XIIIth Century Loggia de Vicari based on structural glass trusses. A case study. In *Proceedings of Glass Processing Days 2005, 17-20 June, Tampere, Finland*. 2005.

[271] Panait, A. *Etude expérimentale et numérique des problèemes de contact unilatéral et de frottement sec dans les assemblages verriers*. Ph.D. thesis, Université de Marne-la-Vallée, France, 2004.

[272] PauliSohn 2007. *Forschungsbericht: Einfluss von Widerstand (Material und Konstruktion) und Einwirkungen auf die Sicherheit von Stahl-Glas-Konstruktionen*. Technical Report, Universität der Bundeswehr München, Auftraggeber Pauli+Sohn GmbH, 2007.

[273] PBS-BSTW. Public Buildings Service – Building Security Technology Website. 2006. URL http://www.oca.gsa.gov/.

[274] Petzold, A., Marusch, H. and Schramm, B. *Der Baustoff Glas: Grundlagen, Eigenschaften, Erzeugnisse, Glasbauelemente, Anwendungen*. 3. Aufl. Verlag Karl Hofmann, Schorndorf, 1990. ISBN 3-7780-1181-2.

[275] Pilkey, W. D. *Peterson's Stress Concentration Factors*. 2nd Edition. Wiley, New York, 1997. ISBN 0471538493.

[276] Pilkington Planar 2005. *Pilkington Planar*. Product documentation, W&W Glass, USA, 2005. URL http://www.wwglass.com/pdf/2005Catalog.pdf.

[277] Pilkington Rec. Design of Glass – Recommendations for the design of glass, -. Pilkington UK Ltd, Undated.

[278] Pilkington TechDoc. Pilkington – Technical documentation. 2006. URL http://www.pilkington.com/.

[279] Porter, M. *Aspects of Structural Design with Glass*. Ph.D. thesis, University of Oxford, 2001.

[280] prEN 13474-1:1999. *Glass in building – Design of glass panes – Part 1: General basis of design*. CEN, 1999.

[281] prEN 13474-2:2000. *Glass in building – Design of glass panes – Part 2: Design for uniformly distributed load*. CEN, 2000.

[282] Pye, A. J. *The Structural Performance of Glass-Adhesive T-Beams*. Ph.D. thesis, University of Bath, 1998.

[283] Quinn, J. B. Extrapolation of Fracture Mirror and Crack-Branch Sizes to Large Dimensions in Biaxial Strength Tests of Glass. *Journal of the American Ceramic Society*, 82(8):2126–2158, 1999.

[284] Raicu, A. and Wilson, H. R. Enhanced Energy Savings in Buildings with Thermotropic Protection Against Overheating. In *EuroSun '96*, pp. 1297–1302. International Solar Energy society, Freiburg, Germany, 1996.

[285] Röder, T. Klebstoffkonstruktionen in Verbindung mit Kfz-Verglasung. In *4. Fachkongress Innovatives Bauen mit Glas*. Bauzentrum München, 1996.

[286] Reed, D. A. and Bradt, R. C. Fracture Mirror-Failure Stress Relations in Weathered and Unweathered Window Glass Panels. *Journal of the American Ceramic Society*, 67(11):C227–C229, 1984.

[287] Reid, S. G. Flaws in the failure prediction model of glass strength. In *Proceedings of the 6th Intemational Conference on Applications of Statistics and Probability in Civil Engineering, Mexico City*, pp. 111–117. 1991.

[288] Reid, S. G. Model Errors in the Failure Prediction Model of Glass Strength. In *8th ASCE Specialty Conference on Probabilistic Mechanics and Structural Reliability, Alberta, Canada*. 2000.

[289] Richter, H. *Experimentelle Untersuchungen zur Rissausbreitung in Spiegelglas im Geschwindigkeitsbereich 10E-3 bis 5 10E3 mm/s*. Ph.D. thesis, Institut für Festkörpermechanik Freiburg, 1974.

[290] Ritter, J. E., Service, T. H. and Guillemet, C. Strength an fatigue parameters for soda-lime glass. *Glass Technology*, 26(6):273–278, December 1985.

[291] Roik, K., Carl, J. and Lindner, J. *Biegetorsionsprobleme gerader dünnwandiger Stäbe*. Ernst & Sohn, Berlin, 1972.

[292] Ruttert, D. *Bestimmung des mechanischen Verhaltens von Klebstoffen im Zugscherversuch unter Kurz- und Langzeitbeanspruchung*. VDI Berichte, Reihe 5, Universität für Bodenkultur, Wien, Austria, 1990.

[293] Saint-Gobain. Saint-Gobain website. 2006. URL http://www.saint-gobain-glass.com/.

[294] Sanderson, K. D., Hurst, S., McKittrick, T., Rim-

mer, D. and Ye, L. Photocatalytic Coatings for Self Cleaning Glass. In *Proceedings of Glass Processing Days 2003, 15-18 June, Tampere, Finland*, pp. 321–325. 2003.

[295] Schlimmer, M. *Zeitabhängiges mechanisches Werkstoffverhalten*. Springer, Berlin, Germany, 1984.

[296] Schmitt, R. W. *Entwicklung eines Prüfverfahrens zur Ermittlung der Biegefestigkeit von Glas und Aspekte der statistischen Behandlung der gewonnenen Messwerte*. Ph.D. thesis, RWTH Aachen, 1987.

[297] Schneider, J. *Festigkeit und Bemessung punktgelagerter Gläser und stossbeanspruchter Gläser*. Ph.D. thesis, TU Darmstadt, Institut für Statik, 2001.

[298] Schneider, J. and Wörner, J. Glass Strength of Annealed and Tempered structural glass in the area of drilled holes. In *Proceedings of Glass Processing Days 2001, 18-21 June, Tampere, Finland*. 2001.

[299] Schober, H. Freeform Glass Structures. In *Proceedings of Glass Processing Days 2003, 15-18 June, Tampere, Finland*, pp. 46–50. 2003.

[300] Scott Norville, H., Harvill, N., Conrath, E. J., Shariat, S. and Mallonee, S. Glass-related injuries in Oklahoma city bombing. *Journal of Performance of Constructed Facilities*, pp. 50–56, May 1999.

[301] SecurityService 2006. Protecting against terrorism. On-line document published by the UK security service MI5, 2006. URL http://www.ukresilience.info/publications/protecting.pdf.

[302] Sedlacek, G., Blank, K., Laufs, W. and Güsgen, J. *Glas im Konstruktiven Ingenieurbau*. Ernst & Sohn, Berlin, 1999. ISBN 3-433-01745-X.

[303] Sglavo, V., Gadotti, M. and Micheletti, T. Cyclic loading behaviour of soda lime silicate glass using indentation cracks. *Fatigue and Fracture of Engineering Materials and Structures*, 20(8):1225–1234, 1997.

[304] Sglavo, V., Prezzi, A. and Zandonella, T. Engineered stress-profile silicate glass: High strength material insensitive to surface defects and fatigue. *Advanced Engineering Materials*, 6(5):344–349, 2004.

[305] Sglavo, V. M. ESP (Engineered Stress Profile) Glass: High Strength Material, Insensitive to Surface Defects. In *Proceedings of Glass Processing Days 2003, 15-18 June, Tampere, Finland*, pp. 74–77. 2003.

[306] Sglavo, V. M. and Green, D. J. Subcritical growth of indentation median cracks in soda-lime-silica glass. *Journal of the American Ceramic Society*, 78(3):650–656, 1995.

[307] Sglavo, V. M. and Green, D. J. Indentation fatigue testing of soda-lime silicate glass. *Journal of Materials Science*, 34(3):579–585, 1999.

[308] Shand, E. B. Experimental Study of Fracture of Glass: II, Experimental Data. *Journal of the American Ceramic Society*, 37(12):559–572, 1954.

[309] Shand, E. B. Breaking Stress of Glass Determined from Dimensions of Fracture Mirrors. *Journal of the American Ceramic Society*, 42(10):474–477, 1959.

[310] Shelby, J. E. *Introduction to glass science and technology*. Cambridge: Royal Society of Chemistry, 1997. ISBN 0-85404-533-3.

[311] Shen, X. *Entwicklung eines Bemessungs- und Sicherheitskonzeptes für den Glasbau*. Ph.D. thesis, Technische Hochschule Darmstadt, 1997.

[312] Shetty, D. K., Rosenfield, A. R. and Duckworth, W. H. Crack Branching in Ceramic Disks Subjected to Biaxial Flexure. *Journal of the American Ceramic Society*, 66(1):C10–C12, 1983.

[313] SIA 260:2003 (e). *SIA 260: Basis of structural design*. Schweizerischer Ingenieur- und Architektenverein, Zürich, 2003.

[314] SIA 261:2003 (e). *SIA 261: Actions on Structures*. Schweizerischer Ingenieur- und Architektenverein, Zürich, 2003.

[315] Siebert, B. *Beitrag zur Berechnung punktgehaltener Gläser*. Ph.D. thesis, Technische Universität Müchen (TUM), 2004.

[316] Siebert, G. *Beitrag zum Einsatz von Glas als tragendes Bauteil im konstruktiven Ingenieurbau*. Ph.D. thesis, Technische Universität München (TUM), 1999.

[317] Siebert G., R. W. New Generation of Point Fixings – Architecture and Safety Aspects. In *Proceedings of Glass Processing Days 2005, 17-20 June, Tampere, Finland*, pp. 530–533. 2005.

[318] Simiu, E., Reed, A., Yancey, C. W. C., Martin, J. W., Hendrickson, E. M., Gonzalez, A. C., Koike, M., Lechner, J. A. and Batts, M. E. *Ring-on-ring tests and load capacity of cladding glass*. NBS Building Science Series 162, U.S. Department of Commerce - National Bureau of Standards, 1984.

[319] Smith, D. Glazing for injury alleviation under blast loading – UK Practice. In *Proceedings of Glass Processing Days 2001, 18-21 June, Tampere, Finland*. 2001.

[320] Smith, P. D. and Rose, T. A. Blast loading and building robustness. *Progress in Structural Engineering and Materials*, 4(2):213–223, 2002.

[321] Sobek, W. and Blandini, L. Konstruktive Verklebungen – Prototyp einer Glasschale, in Transluzente Materialien. pp. 30–32, 2003.

[322] Stamm, K. and Witte, H. *Sandwichkonstruk-*

tionen – Berechnung, Fertigung, Ausführung. Springer Verlag, 1974.

[323] Stavrinidis, B. and Holloway, D. G. Crack healing in Glass. *Physics and Chemistry of Glasses*, 24(1):19–25, 1983.

[324] Szilard, R. *Theory and analysis of plates: classical and numerical methods*. Prentice Hall, 1974. ISBN 0-13-913426-3.

[325] Timoshenko, S. P. and Gere, J. M. *Theory of elastic Stability*. Second Edition. McGraw-Hill, New York, 1961.

[326] Tomozawa, M. Stress corrosion reaction of silica glass and water. *Physics and Chemistry of Glasses*, 39(2):65–69, 1998.

[327] TRAV 2003. *Technische Regeln für die Verwendung von absturzsichernden Verglasungen (TRAV)*. Technical Report, Deutsches Institut für Bautechnik (DIBt), January 2003.

[328] TRLV 2006. *Technische Regeln für die Verwendung von linienförmig gelagerten Verglasungen (TRLV)*. Technical Report, Mitteilungen des Deutschen Instituts für Bautechnik (DIBt), Berlin, 2006.

[329] Ullner, C. *Untersuchungen zum Festigkeitsverhalten von Kalk-Natronsilicatglas nach mechanischer Vorschädigung durch Korundberieselung*. Technical Report, BAM, Berlin, 1993.

[330] Ullner, C. and Höhne, L. *Untersuchungen zum Festigkeitsverhalten und zur Rissalterung von Glas unter dem Einfluss korrosiver Umgebungsbedingungen*. Technical Report, BAM, Berlin, 1993.

[331] Vallabhan, C. V. G. Iterative Analysis of Nonlinear Glass Plates. *Journal of Structural Engineering*, 109(2):489–502, 1983.

[332] Vallabhan, C. V. G., Asik, M. Z. and Kandil, K. Analysis of structural glazing systems. *Computers & Structures*, 65(2):231–239, 1997.

[333] Vallabhan, C. V. G., Das, Y. C. and Ramasamudra, M. Parametric Study of Axisymmetric Circular-Glass Plates. *Journal of Structural Engineering*, 120, No. 5:1663–1671, 1994.

[334] Van Duser, A., Jagota, A. and Bennison, S. Analysis of Glass/Polyvinyl Butyral Laminates Subjected to Uniform Pressure. *Journal of engineering mechanics*, 125:435–442, 1999.

[335] Walker, G. R. and Muir, L. M. An Investigation of the Bending Strength of Glass Louvre Blades. In *Proceedings of the 9th Australian Conference on the Mechanics of Structures and Materials, Sydney, Australia*. 1984.

[336] Weibull, W. A Statistical Theory of the Strength of Materials. *Ingeniors Vetenskaps Akademien (Proceedings of the Royal Swedish Academy of Engineering)*, 151, 1939.

[337] Weibull, W. A Statistical Distribution Function of Wide Applicability. *Journal of Applied Mechanics*, 18:293–297, September 1951.

[338] Wellershoff, F. *Nutzung der Verglasung zur Aussteifung von Gebäudehüllen*. Ph.D. thesis, RWTH Aachen / Shaker Verlag, 2006.

[339] Wellershoff, F. and Sedlacek, G. Stabilization of Building Envelopes with the use of the Glazing. In *Proceedings of Glass Processing Days 2005, 17-20 June, Tampere, Finland*. 2005.

[340] Wiederhorn, S., Dretzke, A. and Rödel, J. Crack growth in soda-lime-silicate glass near the static fatigue limit. *Journal of the American Ceramic Society*, 85(9):2287–2292, 2002.

[341] Wiederhorn, S. and Tornsend, P. Crack healing in Glass. *Journal of the American Ceramic Society*, 53:486–489, 1970.

[342] Wiederhorn, S. M. Influence of Water Vapor on Crack Propagation in Soda-Lime Glass. *Journal of the American Ceramic Society*, 50:407–414, 1967.

[343] Wiederhorn, S. M. and Bolz, L. H. Stress corrosion and static fatigue of glass. *Journal of the American Ceramic Society*, 53(10):543–548, 1970.

[344] Wiederhorn, S. M., Dretzke, A. and Rödel, J. Near the static fatigue limit in glass. *International Journal of Fracture*, 121(1-2):1–7, 2003.

[345] Wigginton, M. *Glass in Architecture*. Phaidon Press Ltd, London, 1996. ISBN 07-148-2922-6.

[346] Wigginton, M. and Harris, J. *Intelligent Skins*. Architectural Press, 2002. ISBN 0750648473.

[347] Wikipedia. Wikipedia – The Free Encyclopedia. 2006. URL http://www.wikipedia.org/.

[348] Wörner, J.-D., Schneider, J. and Fink, A. *Glasbau: Grundlagen, Berechnung, Konstruktion*. Springer Verlag, 2001. ISBN 3-540-66881-0.

[349] www.weibull.com. On-line Reliability Engineering Resources. 2006. URL http://www.weibull.com/.

[350] Young, W. C. and Budynas, R. G. *Formulas for Stress and Strain*. McGraw-Hill Publishing Co., 2002. ISBN 0-071-21059-8.

[351] Zenkert, D. *The Handbook of Sandwich Construction*. Engineering Materials Advisory Service Ltd., United Kingdom, 1997.

[352] Zienkiewicz, O. C. and Taylor, R. L. *The Finite Element Method Set (3 volumes)*. Butterworth-Heinemann, 2005. ISBN 0-750-66431-2.

[353] ZTV-Lsw 88. *Zusätzliche Technische Vorschriften und Richtlinien für die Ausführung von Lärmschutzwänden an Strassen (Additional technical rules for sound screens near roads)*. Mit Änderungen und Ergänzungen 1992, Verkehrsblatt Verlag, Dortmund, 1988.

Index

You may order SED monographs, or get information on other IABSE publications, directly from:

IABSE, ETH Zürich, CH-8093 Zürich, Switzerland
Phone: +41-1-633 2647 Fax: +41-1-633 1241
e-mail: secretariat@iabse.org web: www.iabse.org
Order Form: www.iabse.org/publications/orderform

Structural Engineering Documents 10
Structural Use of Glass, by Matthias Haldimann, Andreas Luible, Mauro Overend

Published: May 2008; 215 pages; ISBN 3-85748-119-2

Structural Engineering Documents 9
Cable Vibrations in Cable-Stayed Bridges, by Elsa de Sa Caetano

Published: October 2007; 188 pages; ISBN 978-3-85748-115-4

Structural Engineering Documents 8
Use and Application of High - Performance Steels for Steel Structures, by Hans-Peter Günther

Published: October 2005; 152 pages; ISBN 3-85748-113-7

Structural Engineering Documents 7
Use of Fibre Reinforced Polymers in Bridge Construction, by Thomas Keller

Published: August 2003; 131 pages; ISBN 3-85748-108-0

Structural Engineering Documents 6
Structural Bearings and Expansion Joints for Bridges, by Günter Ramberger

Published: May 2002; 89 pages; ISBN 3-85748-105-6

Structural Engineering Documents 5 - Revised Version
Introduction to Safety and Reliability of Structures, by Jörg Schneider

SED 5 revised version (2006) includes Variables Processor software, supporting the subjects treated ***free:*** educational version VaP 1.6;
one-year free trial: professional version VaP 2.2

Published: second edition 2006; 109 pages; ISBN 3-85748-093-6